Conversion of English units to SI units.

to convert from	to	multiply by	
		accurate	common
foot (ft)	metre (m)	3.048 000*E − 01	0.305
foot-pound-force (ft · lbf)	joule (J)	1.355 818 E + 00	1.35
foot-pound-force/second (ft · lbf/s)	watt (W)	1.355 818 E + 00	1.35
inch (in)	metre (m)	2.540 000*E − 02	0.0254
gallon (gal US)	metre³ (m³)	3.785 412 E − 03	0.003 78
horsepower (hp)	kilowatt (kw)	7.456 999 E − 01	0.746
mile (mi US Statute)	kilometre (km)	1.609 344*E + 00	1.610
pascal (Pa)†	newton/metre² (N/m²)	1.000 000*E + 00	1
poundal (pdl)	newton (N)	1.382 550 E − 01	0.138
pound-force (lbf avoirdupois)	newton (N)	4.448 222 E + 00	4.49
pound-mass (lbm avoirdupois)	kilogram (kg)	4.535 924 E − 01	0.454
pound-force/foot² (lbf/ft²)	pascal (Pa)	4.788 026 E + 01	47.9
pound-force/inch² (psi)	pascal (Pa)	6.894 757 E + 03	6890
slug	kilogram (kg)	1.459 390 E + 01	14.6
ton (short 2000 lbm)	kilogram (kg)	9.071 847 E + 02	907

* Exact.

† The pascal (Pa) is the new name for the unit of pressure and stress and takes the place of the newton per square metre.

**applied mechanics
of materials**

applied mechanics of materials

JOSEPH EDWARD SHIGLEY

Professor of Mechanical Engineering
The University of Michigan

McGRAW-HILL BOOK COMPANY

New York   St. Louis   San Francisco   Auckland   Düsseldorf   Johannesburg
Kuala Lumpur   London   Mexico   Montreal   New Delhi   Panama   Paris
São Paulo   Singapore   Sydney   Tokyo   Toronto

This book was set in Times Roman by Progressive Typographers.
The editors were B. J. Clark and Michael Gardner;
the designer was Barbara Ellwood;
the production supervisor was Charles Hess.
The drawings were done by Vantage Art, Inc.
Kingsport Press, Inc., was printer and binder.

applied mechanics
of materials

1 2 3 4 5 6 7 8 9 0   KPKP   7 9 8 7 6 5

Library of Congress Cataloging in Publication Data

Shigley, Joseph Edward.
    Applied mechanics of materials.
     1. Mechanics, Applied. 2. Strength of materi-
als. I. Title.
TA350.S354   620.1'12   75-14183
ISBN 0-07-056845-6

contents

preface

The basic concepts of mechanics of materials are few and are easy to explain without using sophisticated mathematical techniques. It is in accordance with this observation that I have tried to write this book so that anyone who is enrolled in a first-year course in mathematics following high-school graduation can read and understand it. Of course, more mathematics is needed beyond the precollege level to adequately develop the theory. To overcome this problem, the needed mathematics is introduced and explained at appropriate places in the text and appendix. Readers already familiar with the needed mathematical concepts can skip these sections, employing them only for reference purposes.

I have also tried to write a *basic* book on the subject—basic in the sense that the book is intended to present all the prerequisite material for courses which may follow, and no more. Mechanics of materials provides a ready vehicle for many interesting mathematical explorations, but these rambles clutter the mind and slow the assimilation of the important concepts. Hopefully, the temptation to adorn the book with such topics has been successfully resisted.

Students are usually more enthusiastic about a subject if they can see and appreciate its value as they proceed into the subject. For this reason, this book is highly design oriented almost from the very beginning. Throughout the book, short practical applications of the theory are included. These will help the reader to understand the usefulness of the theory and to probe deeper into the implications.

Readers need not have studied a course in statics before beginning this book, because statics is introduced directly into the text and the appendix where it is needed. Of course, those who may have studied statics previously can always omit those sections. The statics presented here is sufficient for understanding this book. Other courses may require a knowledge of statics in greater depth, however. For these cases, the usual applied mechanics courses can be used to provide for greater penetration into statics.

Though its use is optional, the modernized metric system of units (SI) has been included together with an adequate selection of problems in each chapter. The pocket electronic calculator is particularly convenient when solving SI problems, and so tables and other material have been arranged to make use of the calculator as convenient as possible.

A book of this kind cannot be structured and put together without the assistance of many persons, especially students. Among those who have been especially helpful to me are Professor Norman C. Harris, The University of Michigan, Ann Arbor, Michigan; Professor Kenneth J. Schneider, California State Polytechnic University, Pomona, California; Professor John J. McNabb, Bradley University, Peoria, Illinois; Professor Alden W. Counsell, Southeastern

Massachusetts University, North Dartmouth, Massachusetts; Professor William S. Hudspeth, Chicago Technical College, Chicago, Illinois; Professor J. Dale Pounds, Purdue University, Lafayette, Indiana; and Mr. B. J. Clark, Engineering Editor, McGraw-Hill Book Company, New York. I am very grateful to these persons for their suggestions, criticisms, and guidance in the difficult task of organizing and assembling the book. And I want to single out Professor Schneider and Professor Counsell to thank for their particularly valuable contributions and suggestions in helping me to polish the final manuscript.

The readers of this book, students and teachers alike, are cordially invited to write me about any errors encountered or suggestions they may have.

JOSEPH EDWARD SHIGLEY

**applied mechanics
of materials**

CHAPTER 1
introduction

1-1. MECHANICS That branch of physics which deals with time and motion and space is called *mechanics. Engineering mechanics* is simply the application of mechanics to the solution of real-life technical problems. Problems involving the study of the trajectory of an interplanetary vehicle, the effect of an earthquake on a building, or the forces that take place when the piston of an automotive engine pushes on its connecting rod are typical problems that utilize engineering mechanics in their solutions. Thus, mechanics is that science which deals with the forces that act upon discrete bodies or elements and the resultant motions of those bodies.

Some of the principal branches of mechanics are statics, dynamics, solid mechanics, and fluid mechanics.

Statics is the study of the forces that act upon bodies when the bodies are at rest or can be assumed to be at rest.

Dynamics is the study of the motions of particles and bodies that result because of the application of forces. The difference between statics and dynamics is that statics is concerned with force systems which act upon motionless bodies, while dynamics is concerned with the motions which result because of the application of force systems.

Solid mechanics deals with the manner in which solid bodies resist the loads or forces that are applied to them. In studying the mechanics of solids, we are dealing mostly with the strength and rigidity (or stiffness) of machines and structures and their parts. This science is used to design and analyze such things as bridges, buildings, airplanes, and trucks, for example, because these things have to be safe, dependable, and long lived.

Fluid mechanics involves the statics and dynamics of fluids. Fluid mechanics is used to study the behavior of a lubricated bearing in an engine, to learn something about the flow of a fluid in a pipe line, or to study the friction of water on the sides of a ship.

1-2. THE FOUNDATIONS OF MECHANICS Engineering mechanics today is still based on the laws pronounced by Sir Isaac Newton, even though the mechanics of Einstein reveals differences. *Time, mass,* and *force* are three fundamental concepts used in mechanics, but we shall not really define them.[1] It is

---

[1] Many volumes have been written by very distinguished scientists concerning these concepts. For a classic, and one of the very best, see Ernst Mach, "The Science of Mechanics," an English translation by Thomas J. McCormack, 9th ed., The Open Court Publishing Co., LaSalle, Ill., 1942.

necessary to discuss these ideas, however, so that we will know what they mean and how they are used.

Time Newton has this to say about *time:*

> Absolute, true and mathematical time, of itself, and by its own nature, flows uniformly on, without regard to anything external.

In Newton's view, an absolutely perfect watch carried on a long and rapid journey through the solar system and back to earth would read the correct time (that is, earth time) upon its return to earth. But the theory of relativity predicts that the watch will be slow in comparison to the "correct time." Thus, the concept of "the correct time" really has no significance.

Mass The concepts of force and mass are inseparable. Newton defined *mass* as: *The quantity of matter of a body as measured by its volume and density.* This is a most unsatisfactory definition because density *is* the mass of a unit volume. We can excuse Newton only by surmising that he perhaps did not really mean it to be a definition. Nevertheless, he recognized the fact that all bodies possess some inherent property that is different than weight. Thus, a moon rock has a certain constant amount of substance, even though its moon weight is different than its earth weight. This constant amount of substance, or quantity of matter, is called the *mass* of the rock. Even on earth, the weight of the rock depends upon where it is measured, because weight is simply the gravitational attraction between the rock and the earth. This attraction varies considerably and depends upon the distance between the mass centers of the two bodies.

We can also think of mass as an inherent property of a body which has something to do with the motions that the body may acquire under various influences. The passenger riding in an automobile in a front–end collision finds himself propelled toward the windshield because of his mass. But if the passenger is sitting in a stationary vehicle that is struck in the rear, then he finds himself pushed against the back of his seat. These effects come about, of course, because of that property of any mass, called *inertia,* which allows it to resist any effort to change its motion.

Of course the mass of a body can be measured by its weight if we have another body whose mass is known. We can then simply compare the respective weights at the same altitude on earth; the unknown mass is related to the known mass by the ratio of the two weights.

Force According to Newton, force is the effort required to impart a certain acceleration to a mass. Our earliest ideas concerning forces, in fact, came about because of our desire to push, lift, or pull various objects. These pushes and pulls are partly used to overcome friction, but are also used to produce motion or to change the motion of objects possessing mass. We intuitively measure the magnitude of a push or a pull by the intensity of the physical exertion required. Our intuitive concept of force includes ideas such as *place of application, direction,* and *magnitude,* which are the characteristics of a force.

1-3. NEWTON'S LAWS   As stated in *Principia,* Newton's three laws are:

> *Law 1:* "Every body perseveres in its state of rest or of uniform motion in a straight line, except in so far as it is compelled to change that state by impressed forces."
> *Law 2:* "Change of motion is proportional to the moving force impressed, and takes place in the direction of the straight line in which such force is impressed."
> *Law 3:* "Reaction is always equal and opposite to action; that is to say, the actions of two bodies upon each other are always equal and directly opposite."

Newton defined *quantity of motion* as the product of mass and velocity, which is called *linear momentum.* Thus, in law 2, by "change of motion" he means change of linear momentum, that is, a change in the quantity of motion. The first two laws can be summarized by the equation

$$\mathbf{F} = km\mathbf{a} \tag{1-1}$$

which is called the *equation of particle motion.* Here, m is the mass of the particle and $\mathbf{F}$ and $\mathbf{a}$ are the force and acceleration, respectively. The equation means that a certain force $\mathbf{F}$ applied in a certain direction to a particle of mass m will impart an acceleration $\mathbf{a}$ to the mass in the same direction as the force. The product $m\mathbf{a}$ is the same as the rate of change of the momentum of the particle. The term k is a positive constant whose value depends upon the units used to express the force, the mass, and the acceleration.

The law expressed by Eq. (1-1) is often called the *law of motion,* which states that "any force $\mathbf{F}$ acting on a particle of mass m will produce an acceleration $\mathbf{a}$ of the particle; both $\mathbf{F}$ and $\mathbf{a}$ have the same sense and the same direction. It is assumed that $\mathbf{F}$ is large enough to overcome all resisting forces."

Newton's third law is called the *law of action and reaction.* For particles, it states that "when two particles react, a pair of interacting forces come into existence which have the same magnitude, but opposite senses, and which act along the straight line common to the two particles."

1-4. STANDARD UNITS OF MEASURE   The International System of Units, called SI in all languages, was standardized by 36 countries (including the United States) at the 11th General Conference on Weights and Measures (CGPM) in 1960. The abbreviation SI comes from the official name: Le Systeme International d'Unites.

The basic unit of length in SI is the *standard metre*[1] (m); the length of the standard metre is equal to

   1 650 763.73*

wavelengths in vacuum of the radiation corresponding to the transition between the levels $2p^{10}$ and $5d^2$ of the krypton 86 atom.

---

[1] The spelling "metre" is used in the English translation of SI to secure worldwide uniformity.
* Spaces are used in SI instead of commas to group numbers in order to avoid confusion with the practice in some European countries of using commas for decimal points.

In the United States, the *yard* (yd) is legally defined by Congress as a 3600/3937 fraction of the standard metre. The *foot* (ft) is one-third of a yard and the *inch* (in) is one-twelfth of a foot.

The *kilogram* (kg) is the basic unit of mass in SI; it is equal to the mass of the international prototype of the kilogram.

In SI the *second* (s) is the basic unit of time. A second is the duration of

$$9\ 192\ 631\ 770$$

periods of the radiation corresponding to the transition between two hyperfine levels of the ground state of the cesium 133 atom.

1-5. GRAVITATIONAL SYSTEMS OF UNITS   The units of measurement used in technology and engineering throughout the world at present are based on Eq. (1-1), with unity used for the constant k. The equation is then rewritten

$$\mathbf{F} = m\mathbf{a} \tag{1-2}$$

where $\mathbf{F}$ is the force acting on a particle of mass m and $\mathbf{a}$ is the acceleration imparted to that mass.

The *weight, W,* of an object is the force exerted upon the object by gravity. Designating the acceleration due to gravity as g, we have

$$W = mg \tag{1-3}$$

Suppose we employ symbols to designate the units to be employed in Eq. (1-2) as follows:

Force, F
Mass, M
Length, L
Time, T

These symbols are to stand for any unit we might choose to employ. For example, L might be taken as inches, feet, miles, millimetres, hands, or rods. Similarly, T could represent seconds, hours, days, or even years. Although these symbols are not numbers, they may be substituted into Eq. (1-2) as if they were; the equality sign then implies that the symbols on one side of the equation are equivalent to those on the other. Acceleration is defined as the rate of change of velocity, and hence it is expressed in units of length divided by time squared. In symbols this would be written L/T^2 or LT^{-2}. So, if we now substitute all of the units into Eq. (1-2), we get

$$F = MLT^{-2} \tag{1-4}$$

In other words, the units of force must be equivalent to the units of mass multiplied by the units of length and divided by the units of time squared. There are a total of four units involved in Eq. (1-4). To satisfy the equality, we are free to choose any three of these as basic units. The fourth will then be a derived unit because it will depend upon the three basic units.

TABLE 1-1
Gravitational systems of units.

| name of system | fundamental units | | | derived unit |
	length	force	time	mass
British foot-pound-second (FPS)	foot (ft)	pound (lb)	second (s)	slug*
Inch-pound-second (IPS)	inch (in)	pound (lb)	second (s)	lb · s²/in
Metre-kilogram-force-second (MKS)	metre (m)	kilogram-force (kgf)	second (s)	kgf · s²/m

* The dimensions of the slug are lb · s²/ft.

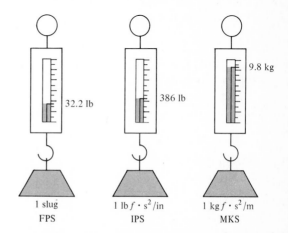

FIGURE 1-1. Scales show the weight of a unit mass under standard gravity for each system of units of the gravitational systems. Standard gravity is roughly the acceleration of gravity at sea level.

A gravitational system of units is one in which the fundamental or basic units are those of *force, length,* and *time.* The mass in a gravitational system is then a derived unit. If we solve Eq. (1-4) for M, we get

$$M = \frac{FT^2}{L} \tag{1-5}$$

which is used to obtain and define the derived unit.

In the English-speaking countries, the British *foot-pound-second* (FPS) and the *inch-pound-second* (IPS) systems are widely used. In the FPS system, the unit of mass is

$$M = \frac{(\text{pound})(\text{second})^2}{\text{foot}} = \text{lb} \cdot \text{s}^2/\text{ft} = \text{slug}$$

This unit of mass has been named a *slug*. In the IPS system, the unit of mass is

$$M = \frac{(\text{pound})(\text{second})^2}{\text{inch}} = \text{lb} \cdot \text{s}^2/\text{in}$$

This unit of mass has not been given another name.

The *metre-kilogram-force-second* (MKS) is used whenever a gravitational system using metric units is desired. Even though 36 countries have already agreed to use SI, the MKS gravitational system is still the favored one among engineers in the so-called metric countries. In the MKS system, the unit of mass is

$$M = \frac{(\text{kilogram-force})(\text{second})^2}{\text{metre}} = \text{kgf} \cdot \text{s}^2/\text{m}$$

All three of the gravitational systems of units are tabulated for your convenience in Table 1-1. Figure 1-1 shows the weight of a unit mass in each of the three systems.

1-6. ABSOLUTE SYSTEMS OF UNITS   Systems of units that are independent of the magnitude of gravitational attraction are called *absolute systems*. In such systems, force is the derived unit. Recalling Eq. (1-4) as

$$F = MLT^{-2}$$

we see that mass, length, and time are the basic units in the absolute systems. Table 1-2 lists the three absolute systems we encounter most frequently in the world literature of engineering, technology, and science.

The unit of force in SI is called the *newton*[1] (N). As indicated in Table 1-2, the newton has dimensions of kilogram-metre per second squared. The name newton was chosen instead of kilogram-force to place additional emphasis on the fact that it is a unit of force and *not* a unit of mass. As shown in the table, the unit of mass in SI is the kilogram.

TABLE 1-2
Absolute systems of units.

name of system	fundamental units			derived unit
	length	*mass*	*time*	*force*
International System of Units (SI)	metre (m)	kilogram (kg)	second (s)	newton (N)*
Centimetre-gram-second (CGS)	centimetre (cm)	gram (g)	second (s)	dyne*
British foot-pound-second (FPS)	foot (ft)	pound-mass (lbm)	second (s)	poundal (pdl)*

* The dimensions of the newton are kg · m/s², of the dyne g · cm/s², and of the poundal lbm · ft/s

[1] A large apple weighs about 1 N.

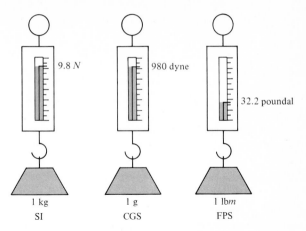

FIGURE 1-2. Scales show the weight of a unit mass under standard gravity for each of the absolute systems of units.

Table 1-2 shows that the *dyne* is the derived unit of force in the CGS system. In the FPS system, the unit of force is called the *poundal.* These two systems are shown only for reference purposes in this book. Figure 1-2 is a comparison of the three absolute systems and shows the weight of a unit mass for each system.

1-7. SYSTEM-TO-SYSTEM CONVERSIONS   Tables A-1, A-2, and A-3 of the appendix will be helpful in converting from or to the International System. Conversions from one system to another are based on Eq. (1-3) in which g is the acceleration due to gravity. The international standard gravity is

$$g = 9.806\ 650 \text{ m/s}^2$$

exactly. This value corresponds roughly to sea level and a 45° latitude. In English units, standard gravity is

$$g = 32.1740 \text{ ft/s}^2$$
$$= 386.088 \text{ in/s}^2$$

For most purposes, the approximate values of 9.80 m/s², 32.2 ft/s², and 386 in/s² are perfectly satisfactory.

In making conversions from one system to another, remember that *weight* is the *force* due to gravity. The terms force, weight, and load, as used in engineering, all have the same meaning. To distinguish them from mass, the terms gram-force (gf), kilogram-force (kgf), and pound-force (lbf) are appropriate. For mass, use the terms gram (g), kilogram (kg), and pound-mass (lbm).

In converting from one system of units to another, always insert the units into the mathematical equation together with the conversion constants. Then treat the units as algebraic terms. For example, the density of concrete in the

FPS absolute system is about 150 lbm/ft³. We convert this to SI units as follows:

$$\rho = \left(150 \frac{\text{lbm}}{\text{ft}^3}\right)\left(0.454 \frac{\text{kg}}{\text{lbm}}\right)\left(3.28 \frac{\text{ft}}{\text{m}}\right)^3 = 2400 \text{ kg/m}^3$$

Note that the conversion constants must be selected so as to cancel the original FPS units.

Few of the so-called metric countries have as yet switched completely to SI. As a result, we find many tabulations of the strengths of various materials in the MKS system. For example, the strength of a certain grade of cast iron is found to be $S = 1410$ kgf/cm². In SI, the prefixes such as milli, mega, etc., cannot be used in the denominator of compound units. This means that we can convert S to newton per square metre (N/m²) or kilonewton per square metre (kN/m²), but that we cannot convert S to newton per square centimetre (N/cm²) nor to newton per square millimetre (N/mm²). Therefore, we convert S as follows:

$$S = \left(1410 \frac{\text{kgf}}{\text{cm}^2}\right)\left(100 \frac{\text{cm}}{\text{m}}\right)^2\left(9.8 \frac{\text{N}}{\text{kgf}}\right)\left(10^{-6} \frac{\text{MN}}{\text{N}}\right) = 138 \text{ MN/m}^2$$

which is read meganewton per square metre. We can convert this to the IPS gravitational system using the conversion constant given in Table A-3 directly. But it can also be converted as follows:

$$S = \left(138 \frac{\text{MN}}{\text{m}^2}\right)\left(\frac{1}{39.4} \frac{\text{m}}{\text{in}}\right)^2\left(10^6 \frac{\text{N}}{\text{MN}}\right)\left(0.225 \frac{\text{lbf}}{\text{N}}\right) = 20\ 000 \text{ psi}$$

where we have used the accepted abbreviation of psi to mean pound-force per square inch.

We have noted above that the prefixes such as mega or milli cannot be used in the denominator of compound units. An exception to this rule occurs with the kilogram (kg) because this is a base unit of SI.

1-8. VECTORS AND SCALARS   A *scalar* is an ordinary real number (positive, negative, or zero) used to express such quantities as time, temperature, or volume. A *vector* is a directed line segment having a magnitude, a direction, and a sense.

The magnitude of a vector is a number proportional to the length of the line segment. The *direction* of a vector is the angle measured from some reference axis to the line containing the line segment. The *sense* of a vector is its orientation positively or negatively along the line, as indicated by its arrow head.

These three characteristics of a vector are illustrated in Fig. 1-3. To determine the sense of a vector, start at the origin and draw a straight line outward at angle θ to x. If the vector points in the direction of this line, it has a positive sense; otherwise, the sense is negative. Note the senses in Figs. 1-3a and c. Both vectors point in the same direction, but the vector in Fig. 1-3c is negative because of the manner in which θ is specified. Compare these two figures with those of Figs. 1-3b and d.

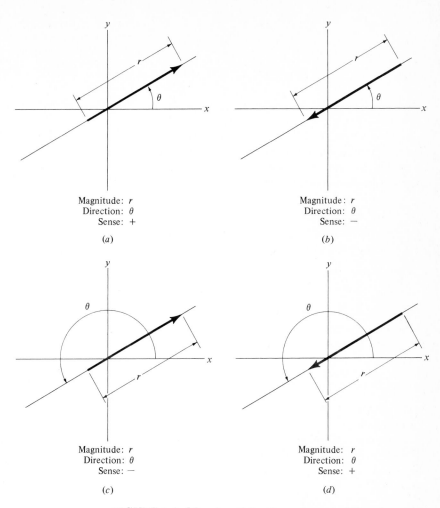

Magnitude: r
Direction: θ
Sense: +

(a)

Magnitude: r
Direction: θ
Sense: −

(b)

Magnitude: r
Direction: θ
Sense: −

(c)

Magnitude: r
Direction: θ
Sense: +

(d)

FIGURE 1-3. Meaning of the three characteristics of a vector.

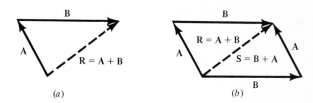

(a)

(b)

FIGURE 1-4. (a) Addition of A and B to form R; (b) the parallelogram rule.

By the phrase *composition of vectors* we mean that vectors may be added or subtracted to produce new vectors. The *vector addition* of **A** and **B** produces a new vector **R** called the *resultant* or *sum*, as shown in Fig. 1-4a. Note that the vectors are composed by placing the tail of **B** at the head of **A**; then **R** extends from the tail of **A** to the head of **B**. But note, in Fig. 1-4b, that the same resultant

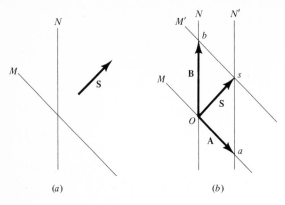

FIGURE 1-5. Resolution of vector S into nonrectangular components A and B.

R is obtained no matter whether you add **B** to **A** or **A** to **B**. Note also that these two sums, **A** + **B** and **B** + **A**, form a parallelogram. It is for this reason that operations involving the composition of vectors are said to utilize the *parallelogram rule*.

To subtract the vector **B** from **A** means that we simply add −**B** to **A**. That is, we take a vector equal and opposite to **B** and add it to **A**.

Suppose now that we wish to take a single vector and from it form two new vectors such that the sum of these new vectors is equal to the original vector. This operation is just the reverse of composition of vectors, and it is called *resolution of vectors*. To illustrate, a vector **S** is given in Fig. 1-5*a* together with lines *M* and *N*. We want to *resolve* the vector **S** into two new vectors **A** and **B** along lines *M* and *N*, respectively, such that

$$\mathbf{A} + \mathbf{B} = \mathbf{S}$$

These two new vectors **A** and **B** are called the nonrectangular component vectors of **S**. As shown in Fig. 1-5*b*, the process is completed using the parallelogram rule.

A particular case of resolution of vectors occurs when the component vectors are at right angles to each other. As shown in Fig. 1-6, the plane vector

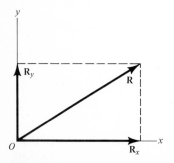

FIGURE 1-6. Rectangular components of R.

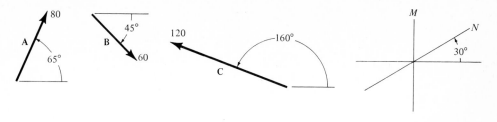

FIGURE 1-7.

R is resolved into components $\mathbf{R}_x$ and $\mathbf{R}_y$ along the *x* and *y* axes, respectively. Since the parallelogram is now a rectangle, the resulting vectors $\mathbf{R}_x$ and $\mathbf{R}_y$ are called the *rectangular components* of **R**.

Notation   A convenient number notation for plane vectors is

$$\mathbf{A} = 16 \angle 40° \qquad \text{or} \qquad \mathbf{B} = 12 \angle -80°$$

where **A** has a magnitude of 16 and is oriented 40° ccw (counterclockwise) from a reference. The ccw direction is *always* taken as the positive direction for the measurement of angles. The vector **B** has a magnitude of 12 and is oriented 80° cw (negative) from the same reference axis. The use of this notation is demonstrated in the example that follows. The symbol $\angle$ means angle.

   In addition, a vector is usually expressed in *boldface* characters. Thus the symbols

   A   **B**   **C**

represent vectors. But it is customary to express the magnitudes of vectors using *lightface* letters. Thus

   A   *B*   *C*

are the magnitudes, respectively, of the vectors **A**, **B**, and **C**.

EXAMPLE 1-1   Figure 1-7 shows the magnitudes and directions of vectors **A**, **B**, and **C**. Carry out the operations specified below using graphical procedures:
   (a) Find **A** + **B**, **A** + **C**, and **B** + **C**.
   (b) Find **A** − **B**, **B** − **A**, and **C** − **A**.
   (c) Resolve **A**, **B**, and **C** into rectangular components in the horizontal and vertical directions.
   (d) Resolve **A**, **B**, and **C** into components along a vertical line *M* and another line *N* pitched 30° above the horizontal.

SOLUTION

   The graphical solutions and answers, shown in Fig. 1-8, should be examined carefully as they are mostly self-explanatory. The magnitudes of the vectors were measured after the graphical work had been completed and the angles were measured using a protractor. In some cases, we have a choice in expressing a vector in numerical notation. In Fig. 1-8d, for example, the vector $\mathbf{C}_N$ could have been written

$$\mathbf{C}_N = -131 \angle 30°$$

////

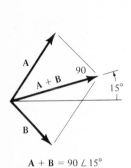

$\mathbf{A} + \mathbf{B} = 90 \angle 15°$

FIGURE 1-8a.

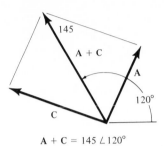

$\mathbf{A} + \mathbf{C} = 145 \angle 120°$

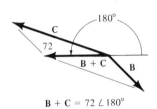

$\mathbf{B} + \mathbf{C} = 72 \angle 180°$

$\mathbf{A} - \mathbf{B} = 116 \angle 94°$

FIGURE 1-8b.

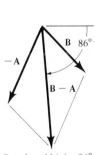

$\mathbf{B} - \mathbf{A} = 116 \angle -86°$

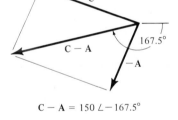

$\mathbf{C} - \mathbf{A} = 150 \angle -167.5°$

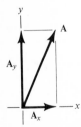

$A_x = 34 \quad A_y = 72.5$

FIGURE 1-8c.

$B_x = 42.5 \quad B_y = -42.5$

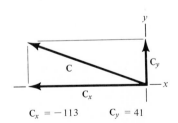

$C_x = -113 \qquad C_y = 41$

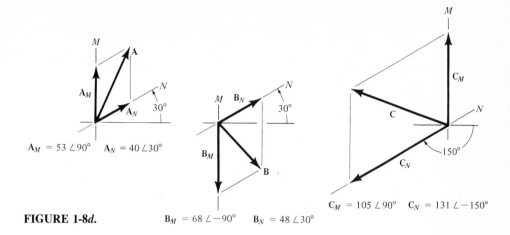

$A_M = 53 \angle 90°$ $A_N = 40 \angle 30°$

$B_M = 68 \angle -90°$ $B_N = 48 \angle 30°$

$C_M = 105 \angle 90°$ $C_N = 131 \angle -150°$

FIGURE 1-8*d*.

1-9. FORCE AND MOMENT   We have already learned that *force* is the effort required to impart an acceleration to a particle of mass. The key words in this definition are force, acceleration, and particle. The characteristics of both force and acceleration are that they have magnitudes, senses, and directions. And a particle is a point. Therefore, force is a concentrated directed quantity and may be represented as a *vector,* as shown in Fig. 1-9. A *force* and a *concentrated force* mean the same thing; they are both force vectors.

In nature, forces may exist because of a gravitational or magnetic field, because of fluid or gas pressures, because of tension in a flexible element such as a rope, belt, or chain, or because of the pushing or pulling action of one body acting against or upon another. These forces always act, or may be considered to act, over a finite area and hence they are called *force fields.* An exception to this rule occurs for gravitational and magnetic forces which, of course, act upon every particle in a body. In practical situations, we can usually sum these forces and consider them to be acting over an area. Figure 1-10*a* is a diagram of a portion of a *pressure force field.* It is a pressure field because the forces act normal to and against the surface.

Figure 1-10*b* is an illustration of a *force distribution.* By this we mean that the forces are distributed or arranged in some continuous manner along a line. A

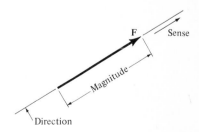

FIGURE 1-9. Force is a vector having a magnitude, a direction, and a sense.

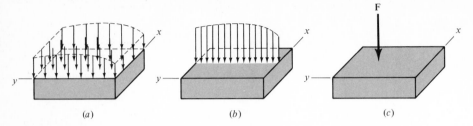

FIGURE 1-10. (*a*) A force field; (*b*) a force distribution; (*c*) a force vector.

force distribution is usually an idealization of a force field. That is, we assume that the force field can be replaced with a force distribution in order to make the analysis easier and the work shorter. In most cases, such idealizations give precisely the same results as would be obtained in analysis by using the force field. A final idealization is obtained by reducing the force distribution to the *force vector* shown in Fig. 1-10*c*.

As indicated earlier, most forces occur in nature as force fields. But we intuitively replace these fields in our own thinking by force vectors. Thus, if we push down on a table with our index finger using, say, a force of 3 lbf, we think of this as a downward acting force vector having a magnitude of 3 lbf. Actually, it is a force field acting throughout the contact area of the finger and table.

Some examples showing how these idealizations or abstractions are made in practical situations are shown in Fig. 1-11. In Fig. 1-11*a* the piston and cylinder head of an automotive engine are subjected to a force field due to the explosion of the fuel-air mixture. As shown in part *b*, this force field may be replaced by the two force vectors F_1 and F_2, which are equal in magnitude, of course, acting on the head and piston.

Figure 1-11*c* shows the probable distribution of forces produced by a wrench in tightening a nut. We might idealize this situation by replacing the force distributions with the force vectors shown in part *d* of the figure.

In Fig. 1-11*e* is shown a force distribution that is an idealization of a pressure field caused by water pushing against a dam. This idealization would be useful if we wanted to analyze the forces acting on a portion of the dam one unit in length.

Finally, in Fig. 1-11*f*, a link *AB* is shown loaded in tension by pins through holes at *A* and *B*. These pins may fit loosely or tightly, we don't know, but the cylindrical sides of the pins push against the contacting surfaces of the holes and create a pressure force field, the exact shape of which is unknown. Therefore, we idealize the situation by replacing each force field by the resultant force vectors **F**, as shown.

Moment The concept of *moment of a force* is illustrated in Fig. 1-12*a*. Here we have given a force *F* and a point *O*. Draw a line through *O* perpendicular to the line of action of **F**. Then the moment of the force is

$$M = hF \tag{1-6}$$

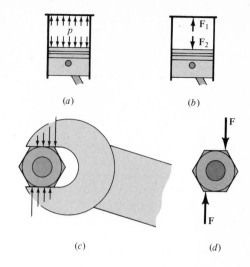

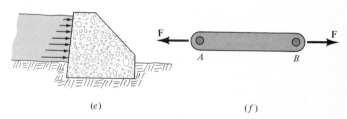

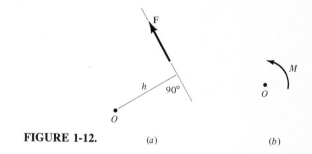

FIGURE 1-11. Some idealizations of force fields and force distributions.

FIGURE 1-12.

where h is the perpendicular distance from O to the line of action of **F** and is called the *moment arm*.

In this book we shall treat the moment M as a scalar quantity and diagram it as shown in Fig. 1-12b. Usually, M is considered positive if acting in the counterclockwise direction.[1]

---

[1] In books on vector mechanics, M is treated as a vector having a direction normal to the plane containing the force and the point O.

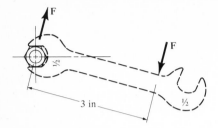

FIGURE 1-13. Definition of a force couple.

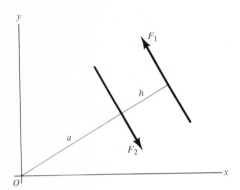

FIGURE 1-14.

Force couples   In Fig. 1-13, a force F is applied to the end of a wrench in tightening a nut. By Newton's law of action and reaction, an equal and opposite force must react on the wrench. This can only occur where the nut makes contact with the wrench. These two forces, the action and the reaction, *are equal and opposite and have parallel lines of action; they cannot be combined to form a single resultant force.* Any two such forces acting on a body constitute a *couple.* The *arm of the couple* is the perpendicular distance between their lines of action. The *plane of the couple* is the plane containing the two lines of action.

The *moment of a couple* is the product of the arm of the couple and the magnitude of one of the forces. Designating the arm by h, we have

$$M = hF \tag{1-7}$$

Force couples have some very interesting properties. To illustrate, in Fig. 1-14 the force pair F_1F_2 form a couple because they are equal, opposite, and parallel. Let us find the moment of these forces about the origin O of the reference system shown. Taking the counterclockwise direction as positive, we have

$$M_O = -aF_2 + (a + h)F_1$$

But $F = F_1 = F_2$, and so

$$M_O = hF \tag{a}$$

which is the same as Eq. (1-7). This demonstration proves:

1   The magnitude of the moment is independent of any particular origin. This means that the moment M can act about any axis so long as that axis is perpendicular to the plane of the couple.

2   The forces of a couple may be rotated together within their plane, keeping their magnitudes and the distance between their lines of action constant, or they may be translated to any parallel plane without changing the magnitude or sense of the moment. Also, two couples are equal if their moments are equal, regardless of the values of the forces or moment arms.

Additional properties are revealed by Fig. 1-15. The force couple F_1F_2 has the lever arm h in Fig. 1-15a, and so the moment is

$$M = hF_1 = hF_2$$

We now wish to use any inclined line of length a between the lines of action to obtain the moment. As indicated by Fig. 1-15b, resolve both forces into components F' and F'' normal to and in line with line a, respectively. The forces F_1'' and F_2'' are equal, opposite, and have the same line of action, and so they cancel each other. Thus, only F_1' and F_2' remain. For these the moment arm is the distance a, and so the moment of this couple is

$$M = aF_1' = aF_2'$$

But, by similar triangles,

$$\frac{h}{a} = \frac{F_1'}{F_1} \qquad \text{or} \qquad F_1' = \frac{h}{a} F_1 \tag{b}$$

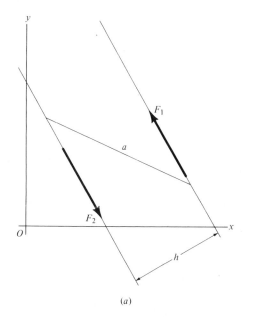

(a)

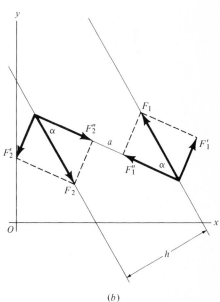

(b)

FIGURE 1-15.

Consequently,

$$M = aF_1' = (a)\left(\frac{h}{a} F_1\right) = hF_1 \tag{c}$$

which is exactly the original moment. This demonstration proves additionally:

  3 The value of the moment M is independent of how the distance between the lines of action is chosen.

Generally, in this book we shall be using the English gravitational systems. This being the case, it is convenient to drop the "f" from the abbreviation "lbf," it being understood that "lb" really stands for pound-force.

PROBLEMS

SECTION 1-5

1-1 A rigid body weighs 50 lb at a place on earth where the acceleration due to gravity is 386 in/s².
  (a) What is the mass of the body in slugs?
  (b) What is the mass of the body in the IPS gravitational system?

1-2 A rigid body weighs 150 lb at a place on earth where the acceleration due to gravity is 32.2 ft/s².
  (a) What is the mass of the body in slugs?
  (b) What is the mass of the body if the IPS gravitational system of units is used?

1-3 A particle of mass experiences an acceleration of 80 in/s² under the action of a force of 4000 lb acting in the same direction as the acceleration. Find the mass of the body using the units of the FPS and IPS gravitational systems.

1-4 A force of 120 lb produces an acceleration of 5 ft/s² when it acts on a certain particle of mass. Find the mass of the particle in both the FPS and IPS gravitational systems of units.

1-5 A body has a mass of 3 lb · s²/in.
  (a) What acceleration would be produced by a force of 80 lb acting on this particle?
  (b) What force would be needed to produce an acceleration of 12 in/s² of this particle?

1-6 A particle has a mass of 3 slugs.
  (a) What force must be applied to give the particle an acceleration of 8 ft/s²?
  (b) What acceleration would the particle acquire if a force of 60 lb were applied to it?

1-7 A force of 70 lb acts on a particle having a mass of 4 lb · s²/in. Find the acceleration of the mass and express the result in in/s² and in ft/s².

1-8 A force of 32 lb acts on a particle having a mass of 12 slugs. Find the acceleration of the mass and express the result in ft/s² and in/s².

1-9 Determine the units of mass in a gravitational system of units where the unit of force is the pound-force, the unit of length is the yard, and the unit of time is the minute. With such units, what acceleration would be produced by a force of 20 lb acting on a mass of 3 units?

1-10 Determine the units of mass in a gravitational system of units where the unit of force is to be in pound-force, the unit of length in miles, and the unit of time in hours. With such units, what force acting on a mass of 8 units would be required to produce an acceleration of 2 miles per hour (mi/h)?

1-11 The accompanying table lists a number of constants used in everyday life in the FPS system of units. Convert these to SI by filling in the blank spaces. Accuracy to three significant figures is sufficient.

no.	name of constant	magnitude in FPS units	magnitude in SI units
a	A person's weight	150 lbm	_____kg
b	Automobile speed	60 mi/h	_____km/h*
c	Gasoline	15 gal	_____m³
d	Lubricating oil	1 pt	_____mm³
e	A 1-acre suburban lot	43,560 ft²	_____m²
f	Atmospheric pressure	14.7 psi	_____kN/m²
g	Density of steel	0.282 lbm/in³	_____Mg/m³

* The correct SI unit is m/s, though speedometers may read in km/h.

Meterologists use a unit of pressure called the *bar* that is associated with the international system, but which is not a part of it. A bar is about one atmosphere and is exactly equal to 100 kN/m².

1-12 Convert the following to the appropriate SI units:
   (a) The area of a city lot that measures 120 ft wide by 200 ft deep.
   (b) The dimensions of a 2 by 10 floor joist. (A 2 by 10 floor joist measures $1\frac{1}{2}$ by $9\frac{1}{4}$ in.)
   (c) The volume of $\frac{1}{2}$ gal of milk.
   (d) The tire pressure corresponding to 30 psi.
   (e) The density of a gallon of paint that weighs 13 lb.

1-13 Convert the following to the most appropriate SI units:
   (a) The density of aluminum from 0.098 lbm/in³.
   (b) The mass of an automobile that weighs 4100 lb.
   (c) The displacement of a 425-in³ automotive engine.*†
   (d) The span of a 1200-ft bridge.
   (e) The hydraulic pressure corresponding to 3000 psi.

* As indicated in Table A-1, prefixes should be used in steps of 1000 in SI. Thus, the use of cm, dm, etc., is not approved. With SI units of higher order, such as m² or mm³, the prefix is also raised to the same power. Thus, mm³ means (mm)³. Under such conditions, the use of nonpreferred prefixes such as cm³ or dm² is permissible.

† The use of the "litre" as a unit of volume is discouraged in SI because it has had two definitions. The litre was originally defined as a cubic decimeter but was later defined as the volume occupied by the mass of 1 kg of pure water at its maximum density under normal atmospheric pressure. This definition produced a volume of 1.000 028 dm³, and in 1964 this definition was withdrawn. Thus the word "litre" is a special name for the cubic decimeter. Its use is permitted in SI, but discouraged.

1-14 The figure shows the magnitudes and directions of a number of vectors.

(a) Using graphical means find the resultants **A** + **B**, **B** + **C**, **C** + **D**, **E** + **F**, **F** + **H**, **A** + **G**, **B** + **G**, **A** + **H**, **D** + **E**, and **B** + **E**.

(b) Find the differences **A** − **B**, **C** − **H**, **G** − **B**, **E** − **D**, **F** − **D**, **A** − **E**, **E** − **A**, **H** − **C**, **B** − **A**, and **D** − **G**.

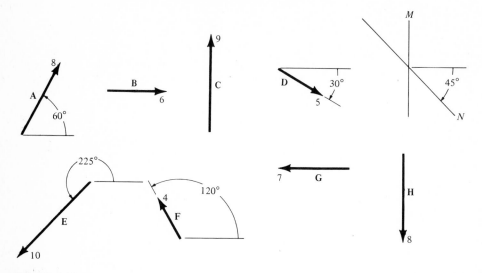

PROB. 1-14.

1-15 We wish to compose the vectors shown in the figure using graphical means.

(a) Find **A** + **B**, **A** + **D**, **B** + **C**, **C** + **E**, **C** + **G**, **D** + **E**, **D** + **F**, **E** + **G**, **B** + **F**, and **C** + **H**.

(b) Find **A** − **C**, **C** − **A**, **B** − **D**, **D** − **B**, **C** − **E**, **D** − **F**, **E** − **F**, **E** − **G**, **H** − **E**, and **H** − **F**.

1-16 Here we wish to resolve the vectors in the figure for Prob. 1-14 using graphical means.

(a) Resolve **A**, **D**, **E**, and **F** into rectangular components in the horizontal and vertical directions.

(b) Resolve vectors **B**, **C**, **D**, **G**, and **H** into components along a vertical line *M* and along a line *N* that dips 45° cw.

1-17 Using the figure for Prob. 1-15, resolve the indicated vectors using graphical means.

(a) Resolve **A**, **B**, **D**, **F**, **G**, and **H** into components in the horizontal and vertical directions.

(b) Resolve vectors **A**, **B**, **C**, **D**, **F**, and **G** into components along a horizontal line *M* and along another line *N* that makes an angle of 60° with *M* in the ccw direction.

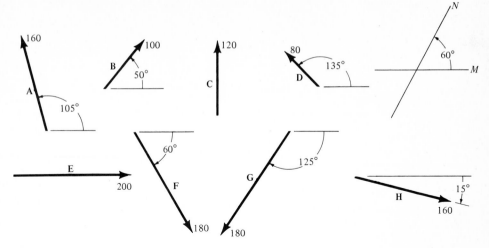

160

100

B
50°

120

C

80

D
135°

N

60°

M

A
105°

E
200

60°

F
180

G
125°

180

H
160

15°

PROB. 1-15.

mechanical properties of materials

2-1. DEFINITION OF STRESS   Stress is a concept invented by man in an attempt to explain what happens within a solid body when that body is pulled, pushed, bent, or twisted, for example, by external effects. As shown in Figs. 2-1 and 2-2, stress is easy to visualize. In each of these figures, a connecting rod is subjected to a push or a pull by a force F acting on pins through the rod ends. In part b of each of the figures, the rod has been cut in two and one part has been removed. We then substitute the stress σ (sigma) on the cut portion as a replacement for the effect of the removed portion. The formula for stress must be, therefore,

$$\sigma = \pm \frac{F}{A} \tag{2-1}$$

where A is the cross-sectional area of the rod. The positive sign is used for tensile stresses and the negative sign for compressive stresses. The units of stress are lb/in^2, which is abbreviated *psi*.

Since $\sigma A = F$, the rod portions shown in Figs. 2-1b and 2-2b obey Newton's law of action and reaction, as might be expected.

In deriving Eq. (2-1), we have assumed that the stress σ is the same for every particle of area on the cut portion of the rod. This is the same as saying that the stress is *uniformly distributed*. To have a uniform stress distribution requires that:

1 The connecting rod be straight.
2 The material be homogeneous.
3 The line of action of the forces be coincident with the centroidal axis[1] of the section.
4 The cross section be remote from the ends of the rod and not close to any abrupt changes in cross section.

The stress defined by Eq. (2-1) is, to be quite specific, properly called *normal stress* because it is normal, that is, perpendicular, to the area of the cross sections. The use of the Greek letter σ is in accordance with scientific notation for stress, though some engineers regularly use the letter S.

Another kind of stress, called *shear stress,* is defined in Fig. 2-3a. Here a portion of a connecting rod is loaded in tension by a force of magnitude $2F$ (as shown) acting on the cut portion of the rod. Another member (not shown) applies counteracting forces F on each end of a pin that projects through the end of the connecting rod. In Fig. 2-3b, the pin has been cut into two portions on one

---

[1] See Appendix A-7 for definition of centroidal axis.

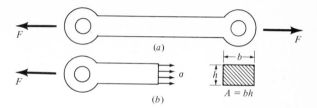

FIGURE 2-1. (*a*) **Connecting rod pulled by forces** *F*; (*b*) **portion of connecting rod showing tensile stress** σ.

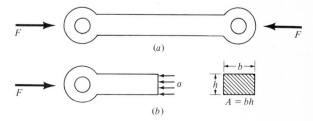

FIGURE 2-2. (*a*) **Connecting rod pushed by forces** *F*; (*b*) **portion of connecting rod showing compressive stress** σ.

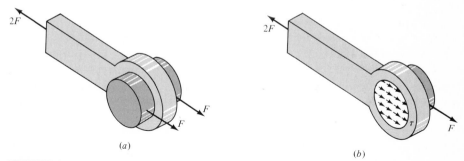

FIGURE 2-3. (*a*) **Connecting rod pulled by tensile force** 2*F*; (*b*) **cross section of pin showing shear stress** τ.

side where it projects from one side of the rod. We then substitute the shear stress τ (tau) on the cut surface as a replacement for the effect of the removed portion. Based on the assumption that this shear stress is uniformly distributed across the cut surface of the pin, the magnitude must be

$$\tau = \frac{F}{A} \qquad (2\text{-}2)$$

where *A* is the cross-sectional area of the pin.

DEFINITION OF STRAIN   Figure 2-4*a* illustrates a connecting rod having ross-sectional area *A* and having scribed upon its body the two marks *A* and *B*

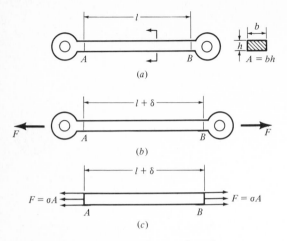

FIGURE 2-4. Definition of normal strain.

a distance l apart. By means of the pins through the ends of the rod, we imagine that tensile forces F are exerted as shown in Fig. 2-4b. Because of the elasticity of the rod, these forces cause the rod to get longer by the amount δ (delta). So the new distance between marks A and B is $l + \delta$. If we like, we can imagine the bar to be cut at each mark and isolated from each end as shown in Fig. 2-4c. In such a case, we must substitute the tensile stress σ at each end to account for the effects of the removed portions. Of course the product of the stress and the area must equal F, at each end, to satisfy Newton's law of action and reaction.

The distance δ is called the *total strain*. If F is a tensile force, δ is a *tensile strain* and is counted as positive. If F is a compressive force, δ is a *compressive strain* and is counted as negative.

The *unit strain* is the strain per unit of original length. It is given by the equation

$$\epsilon = \frac{\delta}{l} \tag{2-3}$$

This strain ϵ (epsilon) is usually measured in inches per inch (in/in), and hence the length l must be in inches. It is called a normal strain because the direction of the strain is the same as the direction of the corresponding normal stress.

Customarily we mean *unit strain* whenever we use the word "strain" without any qualification. It is also customary to use the words *elongation* and *deformation* to mean total strain.

Now let us refer to Fig. 2-5 where we have pictured a round bar of length l. Later in this book we shall discover that the act of twisting the bar sets up a shear stress in the bar. A straight line OA parallel to the axis is first the surface, after which we twist the bar, or imagine that we clockwise direction as shown. This twisting causes A to move t tance e to A'. The *shear strain* is defined as the angle γ (gamma

these particles, say the *j*th one, and find the sum of the forces acting upon it. Let $\mathbf{F}_e$ be the sum of the external forces and $\mathbf{F}_i$ be the sum of the internal forces. Then we can write Eq. (3-1) for the *j*th particle as

$$\Sigma \mathbf{F}_j = \mathbf{F}_e + \mathbf{F}_i = 0 \qquad (a)$$

Now suppose we count all the particles in this system and we find that there are *n* of them all together. For each one of these *n* particles we could write an equation just like Eq. (*a*); after writing them down, we would then have *n* such equations. Now imagine that we add all *n* equations together. We would then have

$$\sum_1^n \mathbf{F}_e + \sum_1^n \mathbf{F}_i = 0 \qquad (b)$$

where the notation

$$\sum_1^n \mathbf{F}$$

means $\mathbf{F}_1 + \mathbf{F}_2 + \mathbf{F}_3 + \cdots + \mathbf{F}_n$, that is, the vector sum of all the forces starting with the first one $\mathbf{F}_1$ and ending with the last one $\mathbf{F}_n$. Consequently, Eq. (*b*) says that a system is in equilibrium if the sum of all the external forces acting on the system plus the sum of the forces acting internal to the system is zero.

Now, Newton's third law of motion, called the law of action and reaction, states that *when two particles react, a pair of interacting forces come into existence, that these forces have the same magnitude, opposite senses, and act along the straight line common to the two particles.* This means that, in the system under study, all the internal forces acting between particles occur in pairs. Each force of the pair, according to Newton's third law, is equal in magnitude and opposite in direction to its mate. Therefore, the second term in Eq. (*b*) is

$$\sum_1^n \mathbf{F}_i = 0 \qquad (c)$$

With only a single term remaining, we have proved that

$$\sum_1^n \mathbf{F}_e = 0 \qquad (3\text{-}2)$$

which states that *the sum of all the external force vectors acting upon a system in equilibrium is zero.*

A procedure quite similar to that used to obtain Eq. (3-2) can be used to demonstrate that *the sum of all the external moment vectors acting upon a system in equilibrium is zero.* Or, in mathematical form,

$$\Sigma \mathbf{M}_e = 0 \qquad (3\text{-}3)$$

Equations (3-2) and (3-3) are quite general and apply to any system no matter how complex it may be. The statements accompanying the equations can also be reversed and it is also true that if *both of these equations are simultaneously satisfied*, then the system is in *static equilibrium*.

For the most part we shall be dealing with forces in a plane or parallel to a plane in this book. This makes it possible to reduce Eqs. (3-2) and (3-3) to a simpler set of relations for our purposes. If we specify that all forces are to be parallel to an xy plane and in the same plane, then Eq. (3-2) can be written in the scalar form

$$\Sigma F_x = 0 \qquad \Sigma F_y = 0 \tag{3-4}$$

and Eq. (3-3) would simply be written as

$$\Sigma M_P = 0 \tag{3-5}$$

since all moments result from forces parallel to the xy plane. The subscript P in this equation identifies the point or axis about which moments are taken. When these two sets of equations are simultaneously satisfied for a body or a system in the xy plane, then the body is in a state of static equilibrium. This means that all the forces in the x direction add to zero, all the forces in the y direction add to zero, and all the moments about any axis normal to the xy plane add to zero.

At the conclusion of the next section, we shall see how these relations are used in actual problems.

3-2. THE CONCEPT OF A FREE-BODY DIAGRAM   In examining the problem of analyzing the behavior, performance, or efficiency of a complex structure or device, such as a bridge, typewriter, or a tractor, the beginner is faced with a bewildering array of complicated parts and geometries. Fortunately, the problem is not as difficult as it seems. One of the most powerful analytical techniques of mechanics is that of isolating or freeing a portion of a system in our imagination in order to study the behavior of one of its segments. When the segment is isolated, the original effect of the system on the segment is replaced by the interacting forces and moments. Figure 3-1 is a symbolic illustration of the process. Let Fig. 3-1a be a total system, such as a bridge. Then we might decide to analyze just one part or segment of the bridge, such as a beam, or even several members joined together. We remove this segment, as in Fig. 3-1b, and

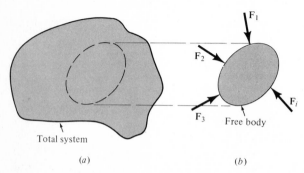

FIGURE 3-1. The isolation of a subsystem.

replace the effect of the whole system on the segment by various forces and moments that would necessarily act at the interfaces of the segment and the system. While these forces may be internal effects upon the whole system, they are external effects when applied to the segment. In the figure, these interface forces are represented symbolically by the force vectors $\mathbf{F}_1$, $\mathbf{F}_2$, $\mathbf{F}_3$, and $\mathbf{F}_i$ in Fig. 3-1b. The isolated subsystem that results, together with all forces and moments due to any external effects and the reactions with the main system, is called a *free-body diagram.*

We can greatly simplify the analysis of a very complex structure or machine by successively isolating each element and studying and analyzing it by the use of free-body diagrams. When all the members have been treated in this manner, the knowledge can be assembled to yield information concerning the behavior of the total system. Thus, free-body diagramming is essentially a means of breaking a complicated problem into manageable segments, of analyzing these simple problems, and then, usually, putting the information together again.

Using free-body diagrams for force analysis serves the following important purposes:

1   The diagram establishes the directions of reference axes, provides a place to record the dimensions of the subsystem, the magnitudes and directions of the known forces, and helps in assuming the directions of unknown forces.

2   The diagram simplifies one's thinking because it provides a place to store one thought while proceeding to the next.

3   The diagram provides a means of communicating your thoughts clearly and unambiguously to other people.

4   Careful and complete construction of the diagram clarifies fuzzy thinking by bringing out various points that are not always apparent in the statement or in the geometry of the total problem. Thus, the diagram aids in understanding all facets of the problem.

5   The diagram helps in the planning of a logical attack on the problem and in setting up the mathematical relations.

6   The diagram helps in recording progress in the solution and in illustrating the methods used.

The next few examples are intended to illustrate the use of free-body diagrams in solving equilibrium problems.

EXAMPLE 3-1   Figure 3-2 illustrates four different devices, various parts of which are to be used to construct free-body diagrams. For each device, we assume there is no friction between contacting parts and, also, that the weight of the various members can be neglected unless it is specified. Though dimensions and forces or loads are given, it is desired only that we do the free-body diagrams of the various parts in this problem without calculating the magnitudes.

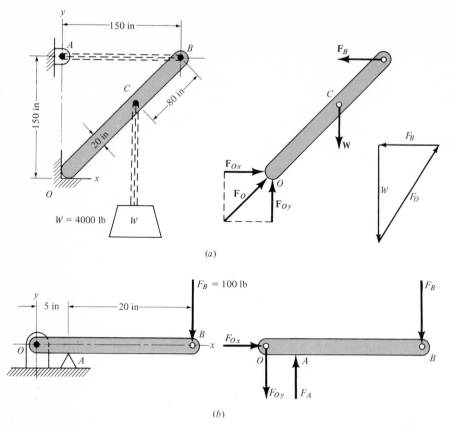

FIGURE 3-2.

SOLUTION

(a) Here we have a rigid member OCB that is loaded by a weight of 4000 lb acting on a pin at C. The member is supported horizontally by a chain pinned at B and is braced against a "corner-stop" at O.

The free-body diagram of the member, shown beside the device, shows how the effects of the contacting members have been replaced by forces. The chain AB exerts a tensile force F_B and so does the chain at C, which is connected to the weight. The member contacts the corner stop at two places and so forces F_{Oy} and F_{Ox} are exerted, respectively, at points on the horizontal and vertical surfaces. Note that these two forces may be summed vectorially to give a resultant $\mathbf{F}_O$.

The member OCB is called a *three-force member*. Whenever the three forces acting on such a system are not parallel, a force triangle, as shown, can be constructed. The force triangle is a graphical representation of Eq. (3-2); it states that the sum of all the forces acting upon a system in equilibrium is zero.

(b) This is a cantilever supported by a hinge at O and a pivot at A, and loaded at B. Since there is no friction, the pivot at A cannot push sideways; thus, F_A must be a vertical force. Writing Eq. (3-4) for the x direction gives $F_{Ox} = 0$. The vertical force F_{Oy} is, therefore, the only force component at O.

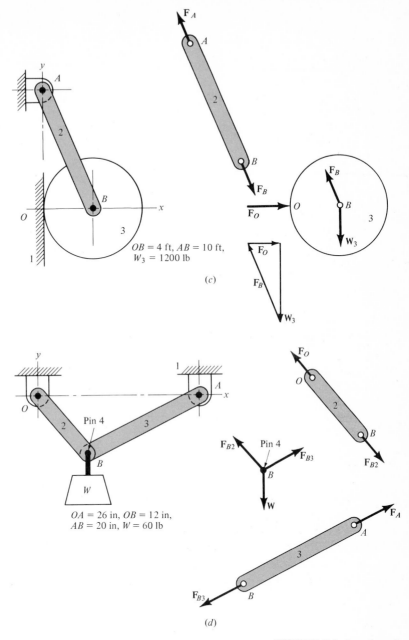

$OB = 4$ ft, $AB = 10$ ft,
$W_3 = 1200$ lb

(c)

$OA = 26$ in, $OB = 12$ in,
$AB = 20$ in, $W = 60$ lb

(d)

FIGURE 3-2. (*continued*)

(c) This device consists of a disk 3 weighing 1200 lb pinned to link 2 and resting against a wall at O.

Link 2 is called a *two-force member*. One force F_A acts through the pin at A, and the force F_B acts through the pin at B. *The two forces of a two-force member always have the same magnitude, their lines of action are always coincident, and the two forces always have opposite senses.* To see that this is true, refer to the free-body diagram of link 2 in Fig. 3-2c. If F_A is not directed along the axis AB, then it will have a moment about point B. But the line of action of F_B is through point B, and so it produces no moment, regardless of its direction. Thus, when we sum the moments about B to zero, only F_A is involved, and its moment arm must be zero to satisfy the equation $\Sigma M_B = 0$. Thus, F_A must act along the line AB. A similar argument can be made for F_B by summing moments to zero about A.

The free-body diagram of disk 3 in Fig. 3-2c consists of two forces in addition to the weight of the disk. The contact force F_O is the force of the wall pushing against the disk. This force is horizontal because there is no friction. Note that the force F_B acting on link 3 is equal and opposite to the force F_B acting on link 2. Since this is a three-force member, a force triangle representing the solution to Eq. (3-2) can be drawn.

The free-body diagram of link 3 also reveals that the three forces F_B, W_3, and F_O are *concurrent*. This means that their lines of action intersect at a common point, B in this case. Provided the forces are not parallel, *any three-force member in static equilibrium will have all three lines of action intersecting at a common point, called the point of concurrency.*

(**d**) In this system, two links of unequal lengths are joined together by pin 4, to which is suspended a weight of 60 lb. To solve the problem, it is necessary to draw free-body diagrams of the pin and of links 2 and 3. As in c, links 2 and 3 can only take axial loads because they are two-force members. Let the pin forces, due to links 2 and 3 acting on the pin, be F_{B2} and F_{B3}, respectively. Then, F_{B2} is in the direction of link 2 and F_{B3} is in the direction of link 3. The remaining results shown are now self-explanatory.   ////

EXAMPLE 3-2   Figure 3-3a depicts a triangular plate, assumed weightless, from which is suspended a weight $W = 800$ lb from a pin at B. The plate is supported on a fixed frame by a hinge at A and by a pin in a vertical slot at O. Draw a free-body diagram of the plate and find all the forces that act upon it.

SOLUTION

The free-body diagram is shown in Fig. 3-3b. Note that the vertical slot in the plate causes the reaction at O to be horizontal. The reaction $\mathbf{F}_A$ at hinge A is unknown in direction. If we attempt a solution to this problem by writing Eq. (3-2), we get

$$\sum_1^3 \mathbf{F}_e = \mathbf{F}_O + \mathbf{F}_A + \mathbf{W} = 0 \tag{1}$$

We now try to solve this equation by graphics in Fig. 3-3c. We then find that we cannot solve it because neither F_O nor F_A have known magnitudes. If we try Eq. (3-4) instead, we get

$$\Sigma F_x = F_O - F_{Ax} = 0 \tag{2}$$

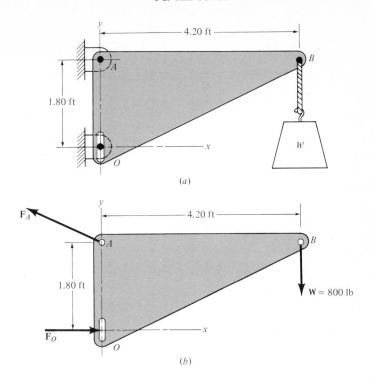

(a)

(b)

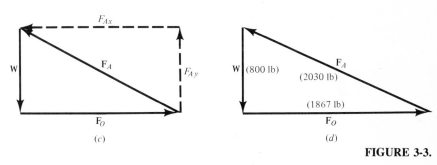

(c)

(d)

FIGURE 3-3.

from which

$$F_O = F_{Ax}$$

and

$$\Sigma F_y = -800 + F_{Ay} = 0 \qquad (3)$$

and so

$$F_{Ay} = 800 \text{ lb} \qquad Ans.$$

Thus, by dividing the force at A into components, we find the vertical component F_{Ay} to be 800 lb. Note particularly that Eq. (1) is a *vector* equation while Eqs. (2)

and (3) are *scalar* equations. Next, since we are unable to find the magnitudes F_O and F_{Ax} by Eq. (3-4), we try Eq. (3-5). Choosing a moment axis through A will make the equation easiest to solve because the perpendicular distances AO to F_O and AB to W are both known. And so, using counterclockwise as the positive direction for moments, we have

$$\Sigma M_A = 1.8\, F_O - (4.20)(800) = 0 \tag{4}$$

from which

$$F_O = \frac{(4.20)(800)}{1.8} = 1867 \text{ lb} \qquad Ans.$$

Since this is a positive result, the assumed direction of F_O is correct.
From Eq. (2) we see that

$$F_{Ax} = F_O = 1867 \text{ lb} \qquad Ans.$$

And if we redraw Fig. 3-3c as Fig. 3-3d using the correct value now found for F_O and scale the result, we find

$$F_A = 2030 \text{ lb} \qquad \theta = 23.2° \qquad Ans.$$

As a practice problem, it is suggested that you take moments about an axis through B to obtain F_O.   ////

EXAMPLE 3-3   A beam loaded by a weight of 120 lb is shown in Fig. 3-4a. Draw a free-body diagram of the beam and compute the unknown reactions. The beam itself weighs 30 lb.

SOLUTION

The free-body diagram is shown in Fig. 3-4b. We next use Eq. (3-5) and take moments about an axis through O. This gives

$$\sum_1^4 M_O = (0)F_O - (3)(30) - (4)(120) + 6F_B = 0$$

where moments are taken as positive in the counterclockwise (ccw) direction. Solving, gives

$$F_B = \frac{(3)(30) + (4)(120)}{6} = 95 \text{ lb} \qquad Ans.$$

With F_B known, it is easy to find F_O using Eq. (3-4). Thus,

$$\sum_1^4 F_y = F_O - 30 - 120 + 95 = 0$$

Note that the y direction is considered as the positive direction for forces. The solution is

$$F_O = 30 + 120 - 95 = 55 \text{ lb} \qquad Ans. \qquad ////$$

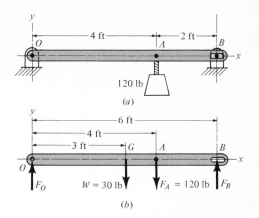

FIGURE 3-4.

3-3. AXIAL STRESSES AND STRAINS In Chap. 2 we learned that an axial load F acting on a member produced a normal stress σ on an area A normal to the direction of F as given by the equation

$$\sigma = \pm \frac{F}{A} \tag{3-6}$$

In this section, we wish to explore the types of structures in which such axial loads occur. It will be easiest to do this by means of examples.

EXAMPLE 3-4 Figure 3-5a shows a structure composed of three members pinned together at their ends. The structure supports a load W; OB is the frame, OA is a horizontal member, and AB is a brace. Assuming frictionless connections and neglecting the weight of the members, determine the loads in OA and in AB.

SOLUTION

We begin by taking the system composed of OA and AB as a free-body; the resulting diagram is shown in Fig. 3-5b. Three external forces act upon this system — the downward force **W** at A, and the frame reactions $\mathbf{F}_O$ at O and $\mathbf{F}_B$ at B. By writing Eq. (3-2), we have

$$\sum_{1}^{3} \mathbf{F}_e = \mathbf{F}_O + \mathbf{F}_B + \mathbf{W} = 0 \tag{1}$$

We can arrange these three force vectors so that their vector sum is zero, as shown in Fig. 3-5c. Note that the triangle formed by $\mathbf{F}_O$, $\mathbf{F}_B$, and **W** is similar to the triangle OAB. This problem can be solved by graphics or by trigonometry. The results are:

$$F_O = \frac{l}{a} W \quad \text{Ans.}$$

$$F_B = \frac{\sqrt{l^2 + a^2}}{a} W \quad \text{Ans.}$$

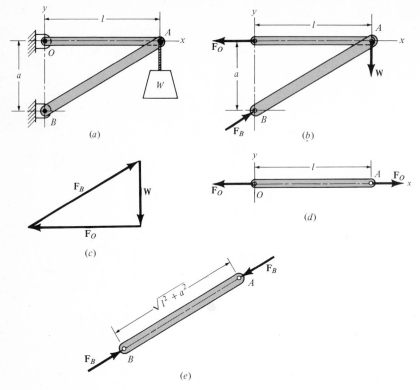

FIGURE 3-5.

Free-body diagrams of the two links are shown in Figs. 3-5*d* and *e*. From these we see that *OA* is in axial tension and *AB* in axial compression. Equation (3-6) can be used to obtain the normal stresses in each member using the negative sign for *AB* because it is in compression.[1]   ////

3-4. SHEAR OF BOLTS, PINS, AND RIVETS   In Sec. 2-1 we learned that the shear stress is given by the equation

$$\tau = \frac{F}{A} \tag{3-7}$$

Figure 3-6 shows two examples in which the use of this equation is appropriate. In *a*, two bars are connected by a single bolt of diameter *d*. Tensile forces *F*

---

[1] A word of caution is necessary concerning the compression member *AB*. A long compression member is more likely to fail by buckling than by exceeding the material strength. In fact, buckling may occur when the actual normal stress is considerably below the yield strength of the material. In Chap. 9 we shall investigate this problem in detail.

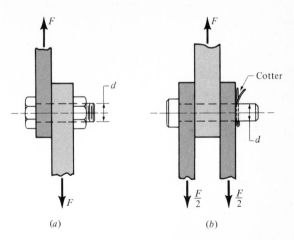

FIGURE 3-6. Shear stresses in bolts and pins. (a) **Single shear of a bolt;** (b) **double shear of a pin.**

acting on the bar tends to shear the bolt across its diameter. The shear stress area is given by the equation

$$A = \frac{\pi d^2}{4} \qquad (a)$$

This is called *single shear* of the bolt.

In Fig. 3-6b, a pin of diameter d connects three parts that are similarly in tension. If the pin were to fail in shear, it would fail in two places. Therefore, the shear area is

$$A = 2\,\frac{\pi d^2}{4} \qquad (b)$$

This is called *double shear*.

The use of Eq. (3-7) implies that a uniform distribution of shear stress exists across the shear area. In actual situations, this is seldom the case and so it is better to call this stress an *average shear stress*. When used for design purposes, a higher factor of safety should be employed to account for these discrepancies.

Of course, Eqs. (a) and (b) are also appropriate when the parts are joined by a rivet, as we shall see in the next section.

3-5. ANALYSIS OF SIMPLE RIVETED JOINTS   The stresses in riveted joints subjected to simple shear loading are usually computed based upon the possibility of three distinct kinds of failure. These are:

1   *Single*, or *double*, *shear* of rivet.
2   *Compression failure* of the rivet, or one of the members, caused by the pressure

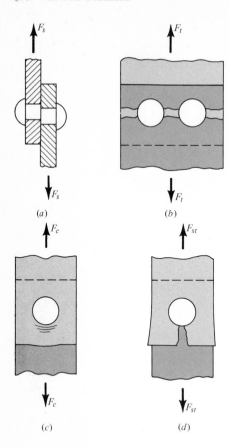

FIGURE 3-7. Modes of failure of a riveted joint. (*a*) Shear of rivet; (*b*) tensile failure of member; (*c*) compression failure of rivet or member; (*d*) shear tearout of member.

between them. This is also called a *bearing failure* because the rivet "bears" on the members.

3 *Tensile failure* of the members between rivet holes.

These three modes of failure together with a fourth, called *shear tearout*, are illustrated in Fig. 3-7. Shear tearout won't occur when the distance from the nearest edge to the rivet center is equal to two or more rivet diameters. Thus, the center of a $\frac{5}{8}$-in rivet should be spaced $1\frac{1}{4}$ in or more from the nearest edge. Since this requirement is usually met, we shall not analyze riveted joints for shear tearout in this book.

One widely used method of analyzing riveted joints is to calculate the allowable load that a joint can carry using appropriate allowable stresses for each assumed mode of failure. We would then use the equations

$$F_s = A_s \tau_{al} \qquad F_b = A_b \sigma_{al} \qquad F_t = A_t \sigma_{al} \tag{3-8}$$

to compute the allowable load F_s that a joint could carry based on shear of the rivets, the allowable load F_b based on compression or bearing of the members or rivets, and the allowable load F_t based on tension of the members. The smallest of these is then taken as the correct allowable load for the joint as a whole. In these equations A_s, A_c, and A_t are the areas subject to shear, compression, and tension, respectively. If the joint is continuous, as in the case of a riveted boiler shell, then the allowable load is computed for a unit length of the joint.

Having computed the allowable load for the joint as a whole, the efficiency can be computed from the equation

$$e = \frac{F_j}{F_{al}} \tag{3-9}$$

where F_j is the allowable load for the joint and F_{al} is the allowable load for the member itself (computed at a section remote from the joint).

Still another approach to the design and analysis of riveted joints is to use the strengths in Eq. (3-8) instead of the allowable stresses. With this change, the equations become

$$F_s = A_s S_s \qquad F_b = A_b S_c \qquad F_t = A_t S_t \tag{3-10}$$

where S_s = shear strength
S_c = compressive strength
S_t = tensile strength

These strengths can be either the yield strengths or the ultimate strengths depending upon the requirements of the problem and whether a construction or design code must be followed or not. The forces obtained from this equation correspond to the forces necessary to develop the entire strength of the joint based upon the three modes of failure. The strength of the joint as a whole would then correspond to the smallest of these. Of course, we would have to apply a factor of safety to the joint strength in order to determine the safe joint load.

When Eq. (3-10) is used to compute the strength of a joint, the efficiency can be computed in a manner quite similar to the previous method. The equation is

$$e = \frac{F_j}{AS} \tag{3-11}$$

where F_j is F_s, F_c, or F_t, whichever is the smallest, A is the area of the member at a section remote from the rivets, and S is the strength of the member.

EXAMPLE 3-5 Figure 3-8 shows a double-riveted lap joint. The allowable tensile stress, shear stress, and bearing stress are, respectively, 20 kpsi, 15 kpsi, and 32 kpsi. Find the allowable joint load and the joint efficiency.

SOLUTION

(a) *Shear of rivets.* Here the rivets are in single shear and we assume that each rivet can take one-half of the joint load. The shear area for a single rivet is

$$A = \frac{\pi d^2}{4} = \frac{\pi (0.375)^2}{4} = 0.110 \text{ in}^2$$

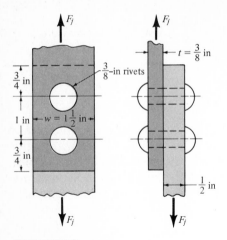

FIGURE 3-8.

where d is the rivet diameter, or the hole size if it is a hot-driven rivet. Using Eq. (3-8) we find

$$F_s = A\,\tau_{al} = (2)(0.110)(15)(10)^3 = 3300\text{ lb}$$

(b) *Bearing.* The bearing or compression area is always taken as the projected area of the portion of the rivet that contacts either member. For a single rivet, this area is the diameter of the rivet times the thickness of the member. But be sure to use the thinnest member if they have different thicknesses. Also, the allowable bearing stress for the member might not be the same as for the rivet. If this is the case, be sure to use the smallest allowable stress of the two. In this example, the bearing area for a single rivet is

$$A = dt = (0.375)(0.375) = 0.141\text{ in}^2$$

where t is the thickness of the thinnest member. Using Eq. (3-8) again, we find

$$F_b = A_b\sigma_{al} = (2)(0.141)(32)(10)^3 = 9000\quad\text{lb}$$

(c) *Tension of member.* The $\tfrac{3}{8}$-in member would fail first on a line through the top rivet. The tensile area is

$$A_t = (w - d)t = (1.5 - 0.375)(0.375) = 0.422\text{ in}^2$$

Therefore, the allowable tensile load is found to be

$$F_t = A_t\sigma_{al} = (0.422)(20)(10)^3 = 8440\text{ lb}$$

Since F_S is the smallest of the three, the allowable joint load is

$$F_j = F_3 = 3300\text{ lb}\qquad Ans.$$

(d) *Efficiency.* Computing the allowable load for the member at a section remote from the rivets gives

$$F_a = A\sigma_{al} = (1.5)(0.375)(20)(10)^3 = 11.25(10)^3\text{ lb}$$

Therefore, the efficiency is

$$e = \frac{F_j}{F_{al}} = \frac{3300}{11.25(10)^3} = 0.293$$

which is quite low and indicates a rather poor joint.   ////

EXAMPLE 3-6   Figure 3-9 shows a section of a continuous single-riveted lap joint. Find the strength in pounds force and the efficiency of a length of the joint equal to the pitch p. The rivets are hot-driven into $\frac{11}{16}$-in diameter holes and the pitch is $1\frac{3}{4}$ in. The plate thickness is $\frac{1}{4}$ in. The ultimate strengths are $S_s = 30$ kpsi, $S_t = S_c = 60$ kpsi for the members, and $S_c = 52$ kpsi for the rivets.

SOLUTION

(a) *Shear strengths of rivets.* We are to compute the strength of the joint based on a section $1\frac{3}{4}$ in long which would then contain one rivet. The rivets are in single shear and so the shear area is

$$A_s = \frac{\pi d^2}{4} = \frac{\pi(11/16)^2}{4} = 0.371 \text{ in}^2$$

Therefore, the strength in pounds force is

$$F_s = A_s S_s = (0.371)(30)(10)^3 = 11.1(10)^3 \text{ lb}$$

(b) *Bearing strength.* Since the compressive strength of the rivet material is less than that of the plates, the rivets govern. The bearing or compression area is

$$A_b = dt = \left(\frac{11}{16}\right)\left(\frac{1}{4}\right) = 0.172 \text{ in}^2$$

And so the bearing strength is

$$F_b = A_b S_c = (0.172)(52)(10)^3 = 8.94(10)^3 \text{ lb}$$

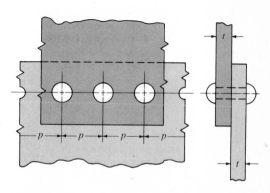

FIGURE 3-9. A single-riveted lap joint. The rivet spacing p is called the *pitch*.

(c) *Tensile strength.* The member would fail in tension along a line through the rivets. The area of a section having a length of one pitch is

$$A = (p - d)t = (1.75 - 0.6875)(0.25) = 0.266 \text{ in}^2$$

Therefore,

$$F_t = A_t S_t = (0.266)(60)(10)^3 = 16(10)^3 \text{ lb}$$

Thus, we find the strength of the joint is governed by the compressive strength of the rivet material and is

$$F_j = 8940 \text{ lb} \qquad Ans.$$

for a length of joint equal to the pitch $1\frac{3}{4}$ in.

(d) *Efficiency.* At a place remote from the rivet, the tensile area of the plate is $A = pt = (1.75)(0.25) = 0.437 \text{ in.}^2$ The efficiency of the joint is

$$e = \frac{F_j}{AS} = \frac{8940}{(0.437)(60)(10)^3} = 0.341 \qquad Ans.$$

It is worth noting that these forces are the strengths and that they must be reduced by a suitable factor of safety to obtain the safe joint load. ////

EXAMPLE 3-7   Figure 3-10 shows a double-riveted butt joint. This connection is $2\frac{1}{2}$ in wide and it is to be loaded in tension. The $\frac{1}{2}$ in rivets are hot driven into $\frac{9}{16}$-in diameter holes. The allowable stresses are 15 kpsi for shear of the rivets, 20 kpsi for tension in the members, and 30 kpsi for compression of the members or rivets. Based on the assumption that the rivets divide the load equally, find the allowable joint load F and the efficiency.

SOLUTION
(a) *Shear of rivets.* The cross-sectional area of a single hot-driven rivet is

$$A = \frac{\pi d^2}{4} = \frac{\pi (0.562)^2}{4} = 0.249 \text{ in}^2$$

The rivets are in double shear and there are three of them on each side of the butt. Therefore, the shear area is

$$A_s = (2)(3)(0.249) = 1.49 \text{ in}^2$$

Then the allowable shear load is

$$F_s = A_s \tau_{al} = (1.49)(15)(10)^3 = 22.4(10)^3 \text{ lb}$$

(b) *Bearing.* The bearing area for a single rivet is

$$A = dt = (0.562)(0.375) = 0.211 \text{ in}^2$$

Note that the bearing area based on the two outside plates would be more than this because they are each $\frac{1}{4}$ in thick. The allowable bearing load is now found as

$$F_b = A_b \sigma_{al} = (3)(0.211)(30)(10)^3 = 19(10)^3 \text{ lb}$$

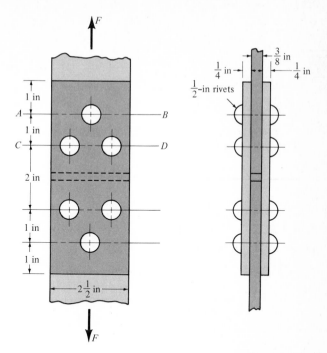

FIGURE 3-10.

(c) *Tension.* In computing the allowable tensile load, we do not really know whether the strength of the joint across the single rivet governs, or across the pair of rivets. So we must try both. Based on line AB through the single rivet, the tensile stress area is

$$A_t = (w - d)t = (2.5 - 0.562)(0.375) = 0.727 \text{ in}^2$$

The allowable tensile load is found to be

$$F_t = A_t\sigma_{al} = (0.727)(20)(10)^3 = 14.5(10)^3 \text{ lb}$$

Next, the tensile stress area on line CD through two rivets is

$$A_t = (w - 2d)t = [2.5 - (2)(0.562)](0.375) = 0.516 \text{ in}^2$$

But the rivet on line AB has already taken out one-third of the load. For this reason, the pair of rivets are subjected to only two-thirds of the total joint load. Therefore, the allowable load is

$$\frac{2}{3}F_t = A_t\sigma_{al}$$

Or

$$F_t = \frac{A_t\sigma_{al}}{2/3} = \frac{(0.516)(20)(10)^3}{0.667} = 15.5(10)^3 \text{ lb}$$

And so we learn that the tensile stress along line *AB* governs. Since this is also the allowable joint load, we write

$$F_j = 14.5(10)^3 \text{ lb} \qquad Ans.$$

(d) *Efficiency.* The area of the member remote from a rivet is

$$A = wt = (2.5)(0.375) = 0.938 \text{ in}^2$$

The corresponding allowable load is

$$F_{al} = AS = (0.938)(20)(10)^3 = 18.8(10)^3 \text{ lb}$$

And so the efficiency is

$$e = \frac{F_j}{F_{al}} = \frac{14.5(10)^3}{18.8(10)^3} = 0.771 \qquad Ans. \qquad \text{////}$$

3-6. BOLTED CONNECTIONS WITH SIMPLE LOADS Rivets are made of relatively ductile materials so that the heads can be cold- or hot-formed. In contrast, bolts should be tough and strong, but not very ductile. The reason for this is that we want to develop the full bolt strength in tightening without very much plastic elongation. A tight bolt is less likely to loosen. Thus, the differences in the analysis of a bolted and a riveted joint come about because a bolt is considerably stronger than a rivet.

Table 3-1 lists the specifications most used for bolts and screws. SAE grades 1 and 2 should only be used for unimportant or non-load-carrying connec-

TABLE 3-1
Material specifications for bolts and screws.

SAE[1] grade	ASTM[2] grade	nominal diameter, in	proof strength, kpsi	tensile strength,[3] kpsi	hardness Bhn	material
1	A307	Up to $1\frac{1}{2}$	33	60	121–241	Low-carbon commercial steel
2		Up to $\frac{3}{4}$	55	74	149–241	Cold-headed, low-carbon steel
		Over $\frac{3}{4}$ to $1\frac{1}{2}$	33	60	121–241	
5	A449	Up to 1	85	120	255–321	Medium-carbon steel; quenched
		Over 1 to $1\frac{1}{2}$	74	105	223–285	and tempered at minimum temperature of 800°F
8	A354	Up to $1\frac{1}{2}$	120	150	302–352	Medium-carbon alloy steel, quenched and tempered to 800°F minimum

[1] Society of Automotive Engineers.
[2] American Society for Testing Materials.
[3] Minimum value.

tions. These materials have a carbon content of 10 or 20 points and are too ductile for loaded connections.

SAE grades 5 and 8 have a carbon content varying from 30 to 45 points, and hence can be heat treated to a high strength with good toughness. These should be used for important load-carrying connections.

The proof strength given in the table is roughly equivalent to the yield strength. The *proof strength* is the maximum tensile stress a bolt can sustain without experiencing a permanent elongation. In nongasketed joints, high-strength bolts should be torqued to create a tensile stress in the bolt at least equal to 90 percent of the proof strength in order to develop the full strength of the bolt.[1]

When a bolted joint is properly tightened and remains tight, the friction between the members is sufficient to carry the shear load. However, there is always the possibility that some loosening may occur, no matter how extensive the precautions taken to prevent it, and it is for this reason that bolted joints are analyzed in a manner quite similar to that for riveted joints. The very same calculations are made: shear failure of the bolt, bearing failure of the bolt or members, and tensile failure of the members. For these modes of failure, the diameter of the bolt used should be the nominal diameter, that is, the size of the unthreaded portion.

In some of the construction codes, bolted joints are analyzed by using the ultimate strengths and proceeding the same as for riveted joints. When this approach is used, one should expect to find the ultimate shear strength of the bolt and the material of the bolt specified by the code used. When this information is not available, a rough guide is to use the ultimate shear strength as equal to 75 percent of the ultimate tensile strength.

But bolted joints are used in a tremendous number of applications that do not require the use of a construction or safety code. In many of these applications, failure is not signified by fracture of the connection but rather by so much permanent deformation or elongation that the connection no longer serves its purpose. This means that yielding should be used as the criterion of failure rather than fracture. For this reason, we shall analyze the connections in this section using the yield strengths.

For bolts we assume, as previously indicated, that the yield strength S_y is the same as the proof strength S_p. Also, we shall take the yield strength in shear as[2]

$$S_{sy} = 0.577S_y \qquad (3\text{-}12)$$

for the bolt material as well as the members. Note that the tensile yield strengths of the members will usually be found in Tables A-16 to A-18. The example that follows is illustrative of the procedure.

---

[1] The explanation for this recommendation is too lengthy and requires more background than we can include here. See Joseph E. Shigley, "Mechanical Engineering Design," 2nd ed., pp. 303–316, McGraw-Hill Book Company, New York, 1972.

[2] For proof of this relation, see Sec. 11-4.

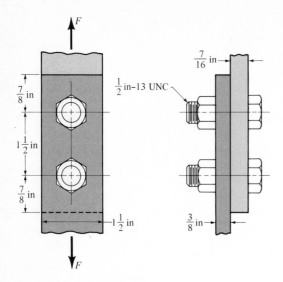

FIGURE 3-11.

EXAMPLE 3-8 Figure 3-11 shows a bolted connection using SAE grade 5 bolts. The members are made of cold drawn AISI 1018 steel. If the tensile load F applied to the connection is 6600 lb, find the factor of safety for all possible modes of failure.

SOLUTION

Entering Table 3-1, we find $S_y = S_p = 85$ kpsi for the tensile and compressive yield strength of the bolt. Employing Eq. (3-13), we find the yield strength of the bolt in shear to be

$$S_{sy} = 0.577(85) = 49.0 \text{ kpsi}$$

Next, we use Table A-18 and find the yield strength in tension and compression for the members to be $S_y = 54$ kpsi. Consequently,

$$S_{sy} = 0.577(54) = 31.2 \text{ kpsi}$$

is the yield strength of the members in shear. For each of the various modes of failure that follow, the factor of safety will be found using either of the equations

$$n = \frac{S_y}{\sigma} \qquad n = \frac{S_{sy}}{\tau}$$

(a) *Shear of the bolts.* Since this is a lap joint, the bolts are in single shear and so the total shear area is

$$A_s = 2 \frac{\pi d^2}{4} = 2 \frac{\pi (0.5)^2}{4} = 0.392 \text{ in}^2$$

Therefore, the average shear stress in the bolts is

$$\tau = \frac{F}{A_s} = \frac{6600}{0.392} = 16.8(10)^3 \text{ psi}$$

And so the factor of safety is

$$n = \frac{S_{sy}}{\tau} = \frac{49.0(10)^3}{16.8(10)^3} = 2.92 \qquad Ans.$$

(b) *Bearing on the bolts.* The $\frac{3}{8}$-in member bears heaviest on the bolts, and so the bearing area is

$$A_b = 2td = 2(0.375)(0.5) = 0.375 \text{ in}^2$$

The corresponding bearing stress is

$$\sigma_b = \frac{F}{A_b} = -\frac{6600}{0.375} = -17.6(10)^3 \text{ psi}$$

which is negative because the stress is compression. The factor of safety is

$$n = \frac{S_{yc}}{\sigma_b} = \frac{-85(10)^3}{-17.6(10)^3} = 4.83 \qquad Ans.$$

(c) *Bearing on the members.* The bearing stress on the members is the same as for the bolts because the areas are identical. Therefore, the factor of safety is

$$n = \frac{S_{yc}}{\sigma_b} = \frac{-54(10)^3}{-17.6(10)^3} = 3.07 \qquad Ans.$$

(d) *Tension of members.* The tensile stress area is

$$A_t = (w - d)t = (1.5 - 0.5)(0.375) = 0.375 \text{ in}^2$$

The tensile stress is

$$\sigma_t = \frac{F}{A_t} = \frac{6600}{0.375} = 17.6(10)^3 \text{ psi}$$

Therefore, the factor of safety is

$$n = \frac{S_{yt}}{\sigma_t} = \frac{54(10)^3}{17.6(10)^3} = 3.07 \qquad Ans. \qquad\qquad ////$$

3-7. ANALYSIS OF BUTT AND FILLET WELDS   Figure 3-12 shows two kinds of arc-welded connections that are subject to relatively simple stresses. The tensile-stress area of the butt joint in part *a* of the figure is

$$A_t = hl$$

And so the tensile stress in the weld metal is

$$\sigma = \frac{F}{hl} \tag{3-13}$$

Note also that the tensile stress in the parent metal, that is, in the members, is the same in this case.

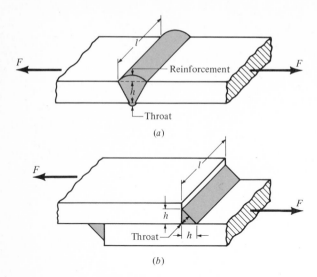

FIGURE 3-12. Two kinds of arc-welded connections.
(a) **Typical butt joint;** (b) **double-filleted lap joint.**

Figure 3-12b shows a double-filleted lap joint. The *leg* of the *fillet* is the distance h. The *throat* of the fillet is $0.707h$. Therefore, the *throat area* for a single fillet is $0.707hl$. An exact solution to the stresses in a fillet weld such as this has never been obtained. It is known to be quite complicated though, and hence designers usually assume a uniform distribution of shear stress across the throat area. Since there are two welds in Fig. 3-12b, the shear stress area is

$$A_s = 2(0.707hl) = 1.414hl$$

Therefore, the *average* shear stress is

$$\tau = \frac{F}{1.414hl} \qquad (3\text{-}14)$$

The AISC specifications for the allowable stresses in the weld metal are 13.6 kpsi for the shear stress on the throat area of a fillet weld, and 13 kpsi on a section through the throat of a butt weld. The specifications also state that the stress in a fillet weld is considered as shear for *any* direction of the applied load.

3-8. THIN-WALL PRESSURE VESSELS   Whenever the wall thickness of a pipe or tank or other pressure vessel is about one-tenth (or less) than its diameter, the stress that results from pressurizing the vessel can be assumed to be uniformly distributed across the wall thickness. Such vessels are called *thin-wall* pressure vessels.

Cylindrical vessels   A cylindrical tank used as a pressure vessel is shown in Fig. 3-13a. The outside diameter is d_o, the inside diameter is d_i, and the wall

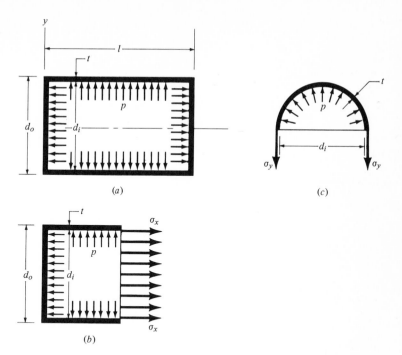

FIGURE 3-13. A cylindrical pressure vessel.

thickness is t. The tank contains a fluid that exerts a pressure of p pounds per square inch (psi) on the inside walls. We want to investigate the stress state in the walls of the tank.

In Figure 3-13b, we see a free-body diagram of about half of the tank. We imagine that we cut the tank in half across its diameter and that we replace the effect of the removed portion on the remaining portion by the uniform tensile stress σ_x acting around the circumference. Thus, the free-body diagram represented by Fig. 3-13b will be in static equilibrium because the force created by the stress acting around the circumference of the cylinder will exactly balance the pressure of the fluid on the circular end wall.

The force developed by the fluid acting against the end wall is

$$F = pA = p\,\frac{\pi d_i^2}{4} = \frac{\pi p d_i^2}{4} \qquad (a)$$

Similarly, the force developed by the stress acting at the cut surface around the circumferential area is

$$F = \sigma_x A = \sigma_x \pi d_i t \qquad (b)$$

Since these two forces are equal and opposite, we can equate them:

$$\sigma_x \pi d_i t = \frac{\pi p d_i^2}{4}$$

Then we solve for σ_x to get

$$\sigma_x = \frac{pd_i}{4t} \tag{3-15}$$

This is called the *longitudinal stress*.

Next, refer to Fig. 3-13c where the vessel has been cut in half in the longitudinal direction. We imagine that the lower half has been removed and we replace its effect on the upper half by the tensile stress σ_y, which is distributed uniformly along the cut edges. The pressure p acting on the inside of the semicircular free-body diagram creates an upward force that must be exactly balanced by the force caused by the stress σ_y acting on the area of the cut edges.

The upward force caused by the pressure p is

$$F = pA = pd_i l \tag{c}$$

The downward force caused by the stress σ_y is

$$F = \sigma_y A = \sigma_y (2tl) \tag{d}$$

Equating these as before gives

$$\sigma_y (2tl) = pd_i l$$

or

$$\sigma_y = \frac{pd_i}{2t} \tag{3-16}$$

This is called the *hoop* or *tangential stress*. Note that it is exactly twice as much as the longitudinal stress.

Frequently, the dimensions of piping are such that they can be treated using the thin-wall assumption too. Pipe carrying fluids under pressure have a beginning and an end, as well as turns, elbows and the like, and so both the hoop and the longitudinal stress components are developed.

This is a good place to bring out the fact that the stress state described by the combination of Eqs. (3-15) and (3-16) is called a *two-dimensional* or *plane-stress situation*. Figure 3-14 illustrates the concept of a stress element.[1] A stress element is a convenient device used to represent a two-dimensional state of stress. Figure 3-14a illustrates the cylinder aligned in the xy coordinate system. A small square of the vessel wall labeled A is selected for examination. We remove this square as a free body in Fig. 3-14b. The stresses σ_x and σ_y acting upon the cut surfaces create forces that maintain equilibrium. This free-body diagram is called a *stress element*. We shall make considerable use of the concept of a stress element in later portions of this book.

Spherical vessels   The spherical vessel of Fig. 3-15a has an inside diameter d_i, a thickness t less than one-tenth d_i, and supports an internal pressure p. Figure 3-15b shows a free-body diagram of half the vessel in which the effect of the

---

[1] See also Sec. 5-6.

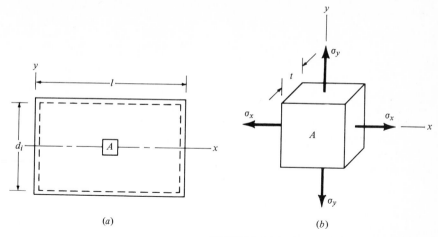

FIGURE 3-14. Concept of a stress element.

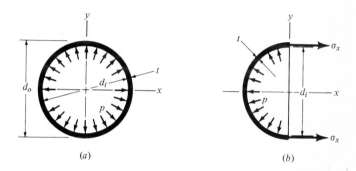

FIGURE 3-15. A spherical pressure vessel.

removed portion has been replaced by the stress σ_x. The pressure p acting to the left creates a leftward acting force whose magnitude is

$$F = pA = p\,\frac{\pi d_i^2}{4} \tag{3}$$

The stress σ_x acting to the right creates another force of magnitude

$$F = \sigma_x A = \sigma_x(\pi d_i t) \tag{f}$$

Since these two forces are equal and opposite,

$$\sigma_x(\pi d_i t) = p\,\frac{\pi d_i^2}{4}$$

Solving for σ_x gives

$$\sigma_x = \frac{p d_i}{4t} \tag{3-17}$$

Had we cut the sphere along the x axis to get the free-body diagram, the same result would have been obtained. Therefore, for a spherical pressure vessel,

$$\sigma_y = \sigma_x \tag{3-18}$$

3-9. STRESS-STRAIN RELATIONS   When two-dimensional stresses are present, as on the stress element of Fig. 3-14b, a change must be made in the equations for strain to account for the Poisson effect. Consider first the *one-dimensional* or *uniaxial* stress element of Fig. 3-16a. The corresponding *strain element* is shown in Fig. 3-16b. Using Eq. (2-6), we find

$$\epsilon_x = \frac{\sigma_x}{E} \tag{3-19}$$

The unit strain in the y direction is found from the definition of Poisson's ratio [Eq. (2-9)] to be

$$\epsilon_y = -\mu\epsilon_x = -\mu\frac{\sigma_x}{E} \tag{3-20}$$

Thus, there is no such thing as a *uniaxial strain element* because a stress in one direction will always result in lateral strains due to the Poisson effect.

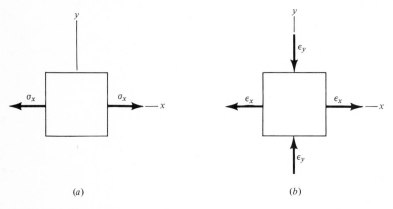

(a)   (b)

FIGURE 3-16.

Figure 3-17a shows a *biaxial* or *two-dimensional stress element*. We can find the corresponding unit strains shown in Fig. 3-17b by the same approach followed previously. The strain ϵ_x is due to σ_x plus the Poisson effect caused by σ_y. Thus,

$$\epsilon_x = \frac{\sigma_x}{E} - \mu\frac{\sigma_y}{E} \tag{3-21}$$

Similarly, for the y direction, we have

$$\epsilon_y = \frac{\sigma_y}{E} - \mu\frac{\sigma_x}{E} \tag{3-22}$$

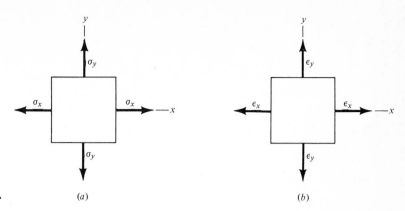

FIGURE 3-17.   (a)   (b)

EXAMPLE 3-9   A steel hoop is to be shrink-fitted to a cast iron wheel 5 ft in diameter in order to serve as a tire. The hoop is $\frac{1}{2}$ in thick, and $3\frac{1}{2}$ in wide, and 0.50 in shorter in circumference than the wheel.
(a) Find the hoop stress
(b) Find the pressure of the hoop on the wheel
(c) Find the biaxial strain

SOLUTION

(a) The circumference of the wheel is

$$C = \pi d = \pi(5)(12) = 188 \text{ in}$$

to three places. Since the elongation of the hoop will be 0.50 in after shrink-fitting, the unit strain in the tangential direction will be

$$\epsilon_t = \frac{0.50}{188} = 2.66(10)^{-3} \text{ in/in}$$

Using Eq. (3-20), we find the hoop or tangential stress to be

$$\sigma_t = E\epsilon_t = 30(10)^6(2.66)(10)^{-3} = 79.8(10)^3 \text{ psi} \qquad Ans.$$

(b) Next, we solve the hoop-stress equation [Eq. (3-16)] for the pressure. This gives

$$p = \frac{2t\sigma}{d_i}$$

Using $d_i = 60$ in and $t = \frac{1}{2}$ in, we get

$$p = \frac{2(0.50)(79.8)(10)^3}{60} = 1330 \text{ psi} \qquad Ans.$$

Note that the width of the hoop doesn't enter into the problem.
(c) We have already found the tangential unit strain to be $\epsilon_t = 2.66(10)^{-3}$ in/in. Using Poisson's ratio $\mu = 0.292$ for steel (see Table A-9), we find the lateral strain to be

$$\epsilon_l = -\mu\epsilon_t = -(0.292)(2.66)(10)^{-3} = -0.777(10)^{-3} \text{ in/in} \qquad Ans. \qquad ////$$

PROBLEMS

SECTION 3-2

3-1 In this problem you are to assume that the parts of the various structures and devices weigh nothing, unless specified. Sketch free-body diagrams of the various parts taking particular care to get the forces in the proper directions. Do not compute magnitudes. Note in each case that the frame is designated as part 1.

(a) Assume W is concentrated at the center of the beam and sketch the forces acting on part 2.

(b) Sketch the free-body diagram of part 2.

(c) Draw the free-body diagrams of the forces acting on the pins.

(d) Draw free-body diagrams of parts 1 and 2.

(e) Sketch a free-body diagram of part 2.

(f) Sketch a free-body diagram of part 2.

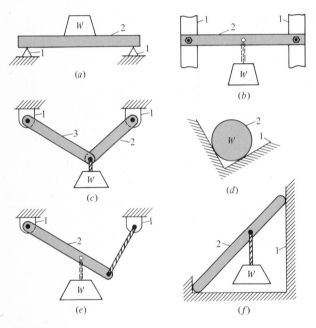

PROB. 3-1.

3-2 Use the instructions in Prob. 3-1 and then:

(a) Draw free-body diagrams of parts 2 and 3.

(b) Draw free-body diagrams of parts 2 and 3.

(c) Draw free-body diagrams of parts 1, 2, and 3.

(d) Draw free-body diagrams of part 1, part 2, and the assembly of parts 3, 4, and 5.

(e) Draw a free-body diagram of part 2.

(f) Draw a free-body diagram of part 2.

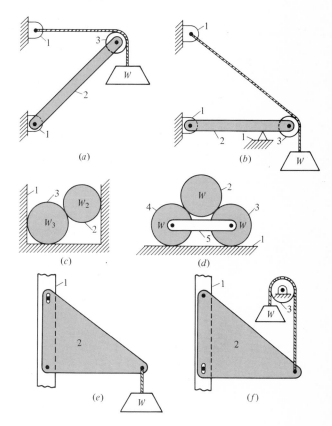

PROB. 3-2.

3-3 The illustration for this problem shows a number of levers actuated by hydraulic cylinders. A hydraulic cylinder pivoted at both ends, as these are, can exert only a push or a pull. Observing the instructions in Prob. 3-1, draw a free-body diagram of each lever.

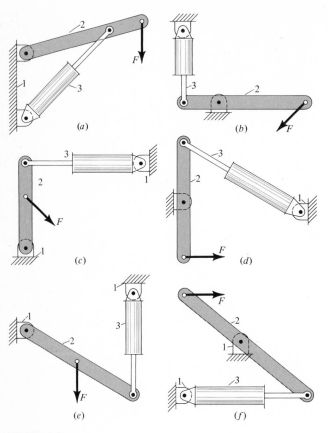

PROB. 3-3.

3-4 Observe the instructions for Prob. 3-1 and draw the following free-body diagrams:
  (a) Parts 3 and 1.
  (b) Parts 3 and 1.
  (c) Part 2.
  (d) Parts 3 and 1.
  (e) Part 3.
  (f) Part 3.

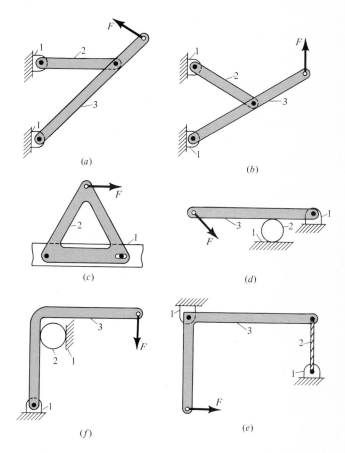

PROB. 3-4.

3-5 Neglecting the weight of the parts, sketch a free-body diagram for part 3 in each case and find the magnitude and direction of the forces which act.

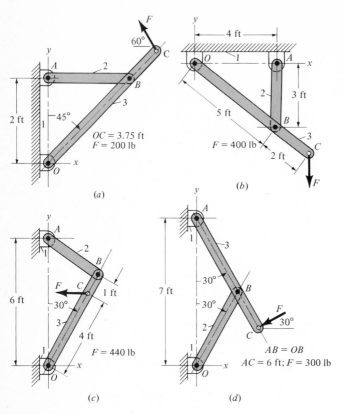

(a)

(b)

(c)

(d)

$OC = 3.75$ ft
$F = 200$ lb

$F = 400$ lb

$F = 440$ lb

$AB = OB$
$AC = 6$ ft; $F = 300$ lb

PROB. 3-5.

3-6 Except where specified, neglect the weight of the parts. Note that the units are in SI.

   (a) Part 2 is a cylinder. Find the direction and magnitude of the forces exerted by the cylinder on frame 1.

   (b) Sketch a free-body diagram of part 2 and find the forces which act.

   (c) Draw a free-body diagram of frame 1 and find both forces.

   (d) Draw free-body diagrams of parts 2 and 4 and compute the magnitude and directions of all forces which act.

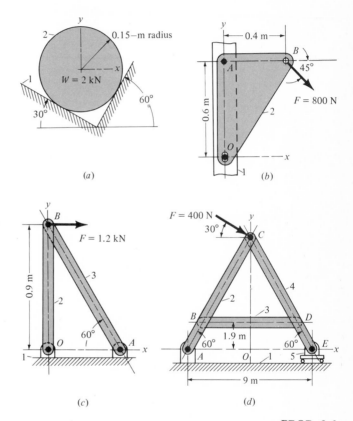

PROB. 3-6.

3-7   Assume the weight of the individual parts can be neglected. In each case, draw a free-body diagram of part 2 and compute the magnitude and direction of all forces exerted on it.

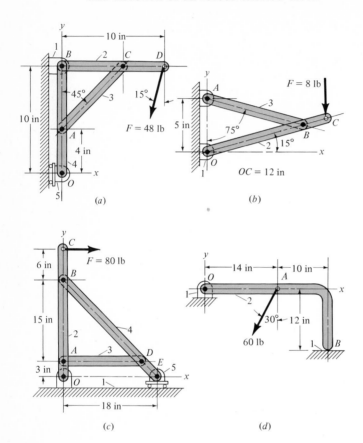

(a)

(b)

(c)

(d)

PROB. 3-7.

3-8 Neglecting the weight of the parts, find the forces which act on part 3 in each case.

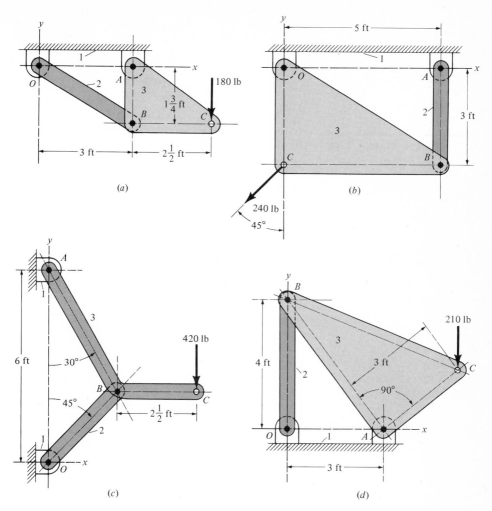

(a)

(b)

240 lb
45°

(c)

(d)

PROB. 3-8.

3-9 The figure illustrates the geometry of a number of beams each of which is loaded by a single concentrated load. Draw a free-body diagram of each beam and find the reactions at the supports. Assume the weight of each beam is small.

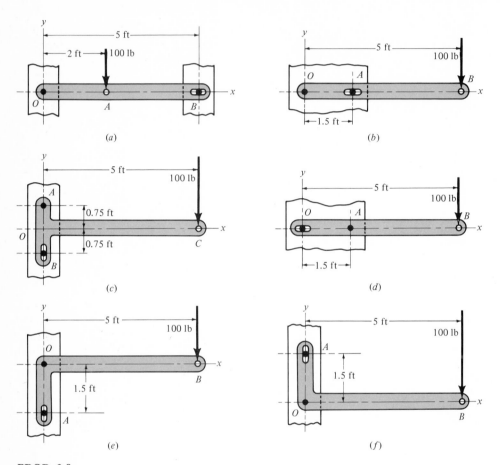

PROB. 3-9.

3-10 Derive formulas for the support reactions for each beam shown in the figure. Neglect the beam weights.

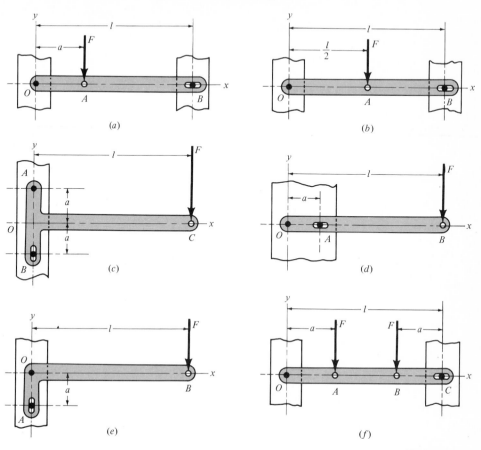

PROB. 3-10.

3-11 The figure displays the free-body diagrams of a number of beams loaded by several external forces. Find all the support reactions R.

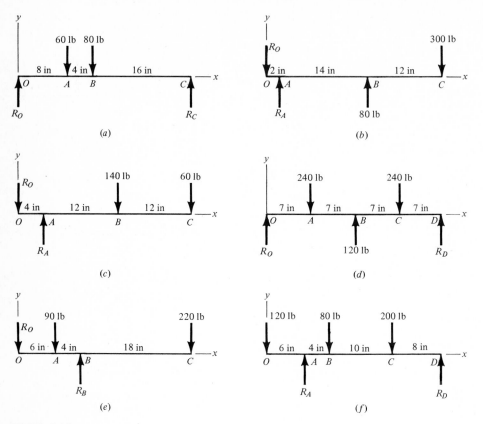

PROB. 3-11.

3-12 The figure shows free-body diagrams of a number of beams. The units are in SI in each case. Find the support reactions R and their correct directions. At least two of the unknown reactions have been assumed in the wrong direction and are so shown on the diagrams.

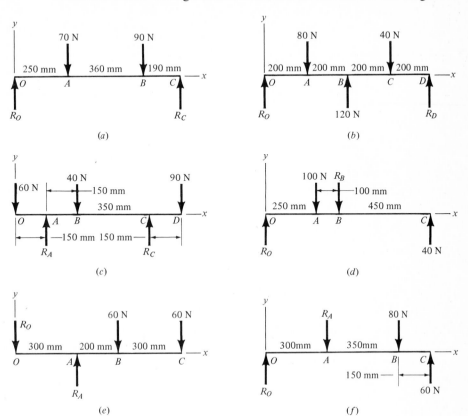

PROB. 3-12.

SECTION 3-3

3-13 A bar is 36 in long and is acted upon by a tensile force of 800 lb. Find the tensile stress and the total elongation of the bar for the following cases:
  (a) The bar is $\frac{3}{8}$ in in diameter and made of steel.
  (b) The bar is $\frac{5}{16}$ in square and is made of ASTM No. 30 cast iron.
  (c) The bar is an aluminum alloy tubing of 1 in outside diameter (OD) by $\frac{1}{8}$ in wall thickness.
  (d) The bar has a rectangular cross section and is made of $\frac{1}{8}$ by $\frac{5}{8}$ in brass.

3-14 A member 16 ft in length is acted upon by a tensile force of 8400 lb. For each of the following cases, find the tensile stress, the total elongation, the unit strain, and the lateral strain:
  (a) The member is a 3 by 3 by $\frac{3}{8}$ in rolled steel angle.

(b) The member is a round glass bar having a $\frac{1}{2}$ in diameter.

(c) The member is a copper tube of 2 in outside diameter (OD) with a wall thickness of $\frac{3}{16}$ in.

3-15 A member has a length of 2.4 m and is acted upon by a tensile force of 12.4 kN. For each of the following cases, find the tensile stress, the total elongation, the axial unit strain, and the lateral unit strain. Use the International System of units.

(a) The member is a round steel bar having a diameter of 6.35 mm.

(b) The member is a rectangular aluminum-alloy bar of 12 by 6 mm in cross section.

(c) The member is a stainless steel tubing having an outside diameter of 20 mm and a wall thickness of 1.5 mm.

3-16 A specimen is 4 in long, 4 in in diameter, and is subjected to a compressive stress of 18 kpsi. Find the total axial contraction and the total lateral expansion of the specimen for the following cases:

(a) The specimen is carbon steel.

(b) The specimen is ASTM No. 20 gray cast iron.

(c) The specimen is concrete.

(d) The specimen is aluminum.

3-17 A round bar 16 in long is to be sized to support a tensile load of 400 lb using a factor of safety of 3. Find an appropriate diameter for the bar for the cases that follow. Use Tables A-16 to A-19 to obtain the strengths. If a range of strengths is given, always use the minimum value in the range.

(a) The material is an aluminum alloy No. 2024-T4.

(b) The material is ASTM No. 25 gray cast iron.

(c) The material is AISI 1018 cold drawn steel.

(d) The material is ASTM No. A306, grade 60, hot-rolled steel.

3-18 Do the same as in Prob. 3-17, except use a factor of safety of 2.7.

(a) The material is ASTM No. A306 grade 50 hot-rolled steel.

(b) The material is AISI 3140 steel heat treated and tempered (drawn) to 800 F.

(c) The material is ASTM No. 40 gray cast iron.

(d) The material is wrought aluminum alloy No. 5056-H18.

3-19 A rectangular bar is to be designed with a width four times the thickness in order to take a compressive load of 1200 lb. The factor of safety is to be 6, and the bar will be braced so as to prevent lateral buckling. Using Tables A-16 to A-19 to get the strengths, find the dimensions of the bar for the following materials:

(a) ASTM No. 25 gray cast iron.

(b) Wrought aluminum alloy No. 1100-H12.

(c) AISI 1010 hot-rolled steel.

(d) AISI 1035 cold-drawn steel.

3-20 Do the same as in Prob. 3-19, except the factor of safety is 5 and the compressive load is 1500 lb.

(a) AISI 1018 hot-rolled steel.

(b) AISI 4340 steel, heat treated and drawn to 1000 F.

(c) Aluminum alloy No. 3004-H36.

(d) ASTM No. 40 gray cast iron.

3-21 An aluminum tube is to be selected to receive a compressive load of 1600 lb. The tube is to be made of alloy No. 3003-H14 using a factor of safety of 6 and an outside diameter of 3 in. Find the largest dimension that can be used safely for the inside diameter.

3-22 A compression member is to be cast as a hollow cylinder 8 in in diameter and 12 in long of ASTM No. 20 gray cast iron. The cylinder is to be subjected to a load of 40 tons. What should be its inside diameter if the factor of safety is to be at least 10?

SECTION 3-5

3-23 The allowable tensile stress, shear stress, and compressive stress for the connections shown in the figure are, respectively, 20 kpsi, 15 kpsi, and 32 kpsi. The rivets are $\frac{1}{2}$ in in diameter and the plate thicknesses t are $\frac{1}{4}$ in. In each case, find the allowable joint load and the efficiency of the connection:

(a) $w = 1\frac{1}{4}$ in

(b) $w = 2$ in

(c) $w = 1\frac{3}{4}$ in

(d) $w = 1\frac{3}{4}$ in

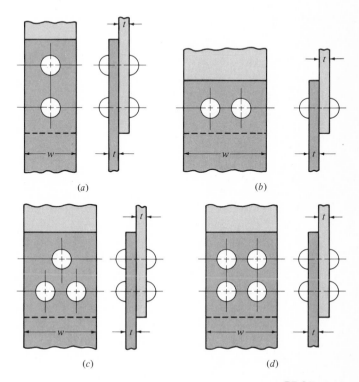

(a) (b)

(c) (d)

PROB. 3-23.

3-24 When ASTM No. A36 steel is used, the AISC specifications for allowable stresses are 16.5 kpsi for shear of rivets, 22 kpsi for tension, and 35 kpsi for bearing stress. Solve Prob. 3-23 using these values instead.

3-25 The connections shown in the figure for Prob. 3-23 are made of materials having ultimate strengths of 33 kpsi for shear of rivets, 56 kpsi for bearing on rivets, and 60 kpsi for the tensile and compressive strengths of the members. The rivets are $\frac{3}{8}$ in in diameter and the plate thicknesses t are $\frac{3}{16}$ in. In each case, find the strength of the joint in pounds force and the joint efficiency.
  (a) $w = 1\frac{1}{8}$ in
  (b) $w = 1\frac{7}{8}$ in
  (c) $w = 1\frac{1}{2}$ in
  (d) $w = 2\frac{1}{8}$ in

3-26 Based upon a factor of safety of 3.5, find the safe tensile load that can be applied to each of the connections shown in the figure for Prob. 3-23. The steel used for the rivets has an ultimate shear strength of 30 kpsi and a compressive strength of 62 kpsi. The steel used for the members has equal tensile and compressive strengths of 70 kpsi. The rivets are $\frac{3}{8}$ in in diameter and are hot driven into holes $\frac{1}{16}$ in larger. The remaining dimensions are as follows:
  (a) $w = 1\frac{7}{8}$ in, $t = \frac{5}{16}$ in
  (b) $w = 3\frac{1}{4}$ in, $t = \frac{3}{8}$ in
  (c) $w = 3\frac{1}{4}$ in, $t = \frac{5}{16}$ in
  (d) $w = 3\frac{1}{2}$ in, $t = \frac{1}{4}$ in

3-27 The allowable tensile stress, shear stress, and compressive stress for the connections shown in the figure are, respectively, 20 kpsi, 15 kpsi, and 32 kpsi. The $\frac{1}{2}$-in rivets are hot driven into $\frac{9}{16}$-in diameter holes, and the plate thicknesses are $t_1 = \frac{3}{16}$ in and $t_2 = \frac{5}{16}$ in. For each case, find the allowable joint load and the efficiency of the connection.
  (a) $w = 1\frac{1}{4}$ in
  (b) $w = 1\frac{1}{4}$ in
  (c) $w = 1\frac{1}{4}$ in
  (d) $w = 2\frac{1}{8}$ in

3-28 The connections shown in the figure for Prob. 3-27 are made of materials having ultimate strengths of 33 kpsi for shear of the rivets, 56 kpsi for bearing on the rivets, and 60 kpsi for the tensile and compressive strengths of the members. The rivets are $\frac{3}{8}$ in in diameter and the plate thicknesses are $t_1 = \frac{3}{16}$ in and $t_2 = \frac{5}{16}$ in. In each case, find the strength of the joint in pounds force and the efficiency.
  (a) $w = 1\frac{1}{4}$ in
  (b) $w = 1\frac{1}{8}$ in
  (c) $w = 1\frac{3}{8}$ in
  (d) $w = 2\frac{1}{8}$ in

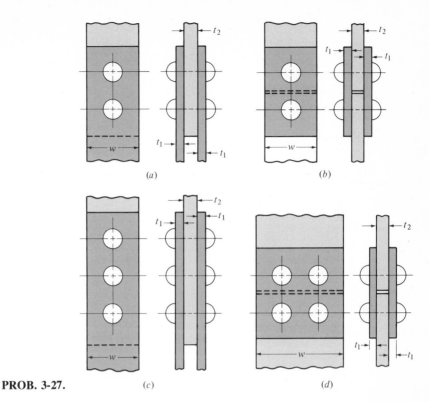

(a)　　　　　(b)

PROB. 3-27.　　(c)　　　　　(d)

3-29 Based upon a factor of safety of 3 for all modes of possible failure, find the safe tensile load that can be applied to each of the connections shown in the figure for Prob. 3-27. The ultimate strength of the steel used for the rivets is 62 kpsi in compression and 46 kpsi in shear. The steel used for the members has an ultimate strength in tension and compression of 70 kpsi. The rivets are $\frac{5}{8}$ in in diameter and are hot driven into holes $\frac{11}{16}$ in in diameter. The remaining dimensions are as follows:

(a) $w = 1\frac{7}{8}$ in, $t_1 = \frac{1}{4}$ in, $t_2 = \frac{5}{8}$ in

(b) $w = 2$ in, $t_1 = \frac{5}{16}$ in, $t_2 = \frac{7}{16}$ in

(c) $w = 1\frac{3}{4}$ in, $t_1 = \frac{5}{16}$ in, $t_2 = \frac{5}{8}$ in

(d) $w = 3\frac{1}{4}$ in, $t_1 = \frac{1}{4}$ in, $t_2 = \frac{1}{2}$ in

3-30 When ASTM No. A36 steel is used, the AISC specifications for the allowable stresses are 16.5 kpsi for shear of rivets, 22 kpsi for tension of members, and 35 kpsi for bearing stress. Using the figure for Prob. 3-27, take the rivet diameters as $\frac{1}{2}$ in, $t_1 = \frac{1}{4}$ in, and $t_2 = \frac{7}{16}$ in and compute the allowable joint loads and efficiencies for each case. The remaining dimensions are:

(a) $w = 1\frac{1}{2}$ in

(b) $w = 1\frac{3}{4}$ in

(c) $w = 1\frac{1}{2}$ in

(d) $w = 3\frac{3}{8}$ in

SECTION 3-6

3-31 The figure shows a bolted lap joint that uses SAE grade 8 bolts. The members are made of cold-drawn AISI 1040 steel. Find the safe tensile load F that can be applied to this connection if the following factors of safety are specified: shear of bolts 3, bearing on bolts 2, bearing on members 2.5, and tension of members 3.

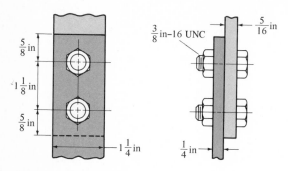

PROB. 3-31.

3-32 The bolted connection shown in the figure uses SAE grade 5 bolts. The members are hot-rolled AISI 1018 steel. A tensile load $F = 4000$ lb is applied to the connection. Find the factor of safety for all possible modes of failure.

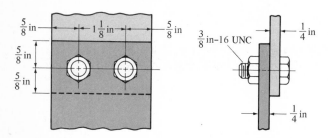

PROB. 3-32.

3-33 A bolted lap joint using SAE grade 5 bolts and members made of AISI 1035 steel, heat treated and drawn at 1000 F, is shown in the figure. Find the tensile load F that can be applied to this connection if the following factors of safety are specified: shear of bolts 1.8, bearing on bolts 2.2, bearing on members 2.4, and tension of members 2.6.

3-34 The bolted connection shown in the figure is subjected to a tensile load of $20(10)^3$ lb. The bolts are SAE grade 5 and the material is cold-drawn AISI 1015 steel. Find the factor of safety of the connection for all possible modes of failure.

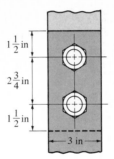

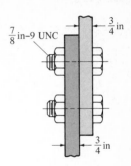

PROB. 3-33.

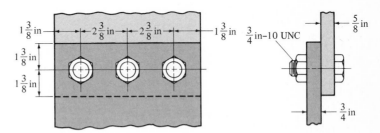

PROB. 3-34.

3-35 The figure shows a connection which employs three SAE grade 5 bolts. The tensile load on the joint is 5400 lb. The members are cold-drawn bars of AISI 1112 steel. Find the factors of safety for each possible mode of failure.

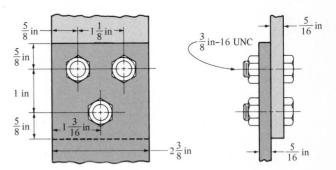

PROB. 3-35.

3-36 The external load to be applied to the joint shown is $16(10)^3$ lb. The bolts are SAE grade 5 and the members are AISI 1018 cold-drawn steel. What is the factor of safety for the connection?

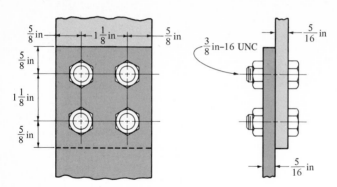

PROB. 3-36.

3-37 An external tensile load of $30(10)^3$ lb is to be applied to the bolted lap joint shown in the figure. The bolts are SAE grade 8, and the members are made of AISI 2330 nickel steel bars that have been heat-treated and drawn at 1000 F. What is the factor of safety of the connection?

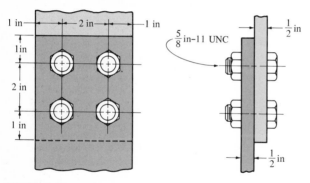

PROB. 3-37.

3-38 SAE grade 8 bolts are used in the joint shown in the figure. The plates are cold-drawn AISI 1035 steel. What safe tensile load F will the connection carry if the factor of safety is to be at least 3 for each possible mode of failure?

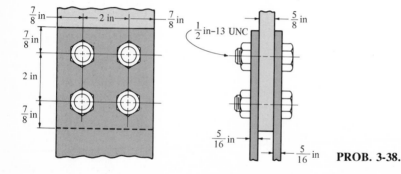

PROB. 3-38.

SECTION 3-7

3-39   Based upon the AISC specifications, find the safe external tensile load that can be applied to each of the joints shown in the figure.

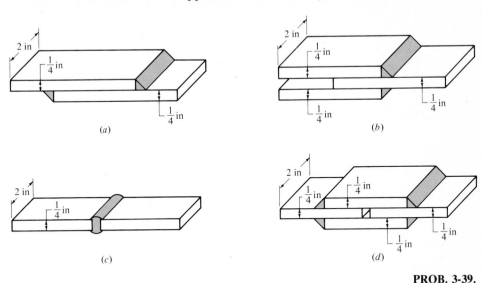

PROB. 3-39.

3-40   Find the significant or critical stress for each weld in the joints for Prob. 3-39. In each case, the external tensile load is 4000 lb.

3-41   The figure shows a collection of welded joints in which the sizes and types of welds are indicated by standardized welding symbols. A small triangle indicates a fillet weld. The dimension preceding the triangle is the size of the leg. If the triangle is on the bottom of the symbol, the arrow shows where the fillet is to be welded; this is called the *near side*. If the triangle is on top of the symbol, then the fillet is to be welded on the *side opposite* of where the arrow points; this is called the *far side*. Of course, a triangle on both sides of the symbol indicates that the weld should be placed on both the near and the far sides. In this problem you are to use the AISC specifications to find the allowable load F that can be applied to each connection shown.

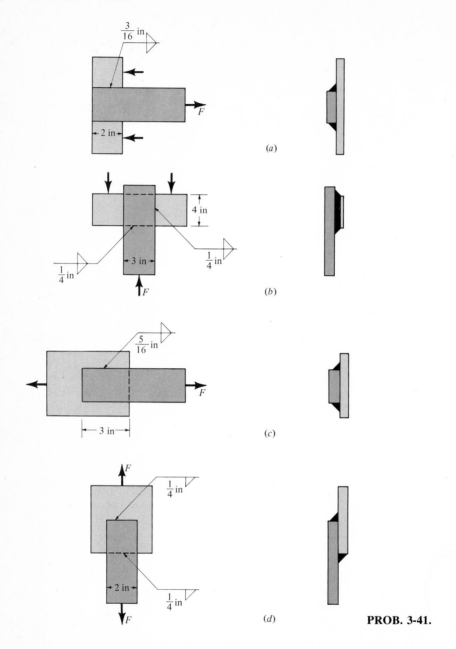

(a)

(b)

(c)

(d)   **PROB. 3-41.**

SECTION 3-9

3-42   A steel hoop 36 in in diameter with a cross section 2 in wide and $\frac{1}{4}$ in
thick has been shrink-fitted over a cast-iron wheel.

(a) What pressure will the hoop exert on the wheel if a hoop stress
of 30 kpsi is permitted?

(b) How much does the circumference of the hoop change because
of the shrink fit?

3-43 A steel hoop is to be shrunk onto a wheel 42 in in diameter. The hoop has a cross section $2\frac{1}{2}$ in in width by $\frac{3}{8}$ in in thickness. If the circumference of the hoop is 0.200 in less than that of the wheel prior to shrink-fitting, what hoop stress results?

3-44 Steel pipe is manufactured in three weights — standard, extra strong, and double extra strong. An approximate permissible pipe pressure can be found by assuming that the thin-wall theory holds for pipe. Use an allowable tensile stress of 9000 psi and compute the permissible pipe pressure for the following sizes:
(a) 4 in standard, 4.500 in OD by 4.026 in ID.[1]
(b) 4-in extra strong, 4.500 in OD by 3.826 in ID.
(c) 4-in double extra strong, 4.500 in OD by 3.152 in ID.

3-45 The same as Prob. 3-44 for the following pipe sizes:
(a) 2-in standard, 2.375 in OD by 2.067 in ID.
(b) 2-in extra strong, 2.375 in OD by 1.939 in ID.
(c) 2-in double extra strong, 2.375 in OD by 1.503 in ID.

SI UNITS

3-46 In the figure for Prob. 3-23, the rivets are 10 mm in diameter and the plate thicknesses t are 6 mm. The allowable tensile stress, shear stress, and compressive stress for the connections are, respectively, 200 MPa, 150 MPa, and 320 MPa. In each case, find the permissible joint load and the efficiency:
(a) $w = 30$ mm
(b) $w = 50$ mm
(c) $w = 45$ mm
(d) $w = 45$ mm

3-47 The figure shows a connection that employs three metric bolts which have a proof strength of 650 MPa. The tensile load on the joint is 25 kN. The members are cold-drawn bars having a yield strength of 425 MPa. Find the factor of safety for each possible mode of failure.

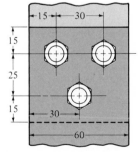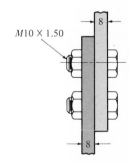

PROB. 3-47. All dimensions in mm. The designation M10 $\times$ 1.50 refers to metric screw threads; it means that the nominal screw diameter is 10 mm and that the pitch, that is, the distance from thread to thread, is 1.50 mm.

---

[1] OD means outside diameter and ID inside diameter.

loading, shear force, and bending moment in beams

4-1. DEFINITIONS AND CONVENTIONS   A *beam* is any structural member that bends when it is subjected to forces perpendicular to its axis. Figure 4-1 illustrates three examples of beams. A rotating shaft supporting two pulleys is shown in Fig. 4-1a. The shaft is supported in bearings at O and C and is bent by the belt pulls P_1, P_2, P_3, and P_4. Section CD of the shaft *overhangs* its bearing, and this section could also be bent by connecting something else, a gear for example, to the shaft.

Figure 4-1b is typical of some of the structural steelwork found in modern skyscrapers. The column, which is usually a *wide flange* or an *H section,* extends from the foundation to the top of the building. Floor beams are then fastened between pairs of columns at intervals corresponding to the distance between floors. These beams are generally made from wide-flange sections or from *I-beam* sections. Note that the beam is riveted to the column using four *angles*. Two of these are riveted to the top and bottom flanges of the beam and the other two are riveted to the web. Sometimes the web connections alone are sufficient to transfer the floor load into the column.

Figure 4-1c shows the worm shaft of a worm-gear speed reducer. Bending of this shaft is produced by the force of the worm gear (not shown) acting against the worm. In addition to the bending loads, this shaft is subjected to *axial* or *thrust loads* because of the helical shape of the gear teeth. Note particularly that the left-hand ball bearing is mounted in its housing so that it can transfer both radial and thrust loads into the frame. The right-hand ball bearing is called a *floating bearing* because it can slide in the axial direction in its housing. This means that it can transfer radial loads to the frame, but not thrust loads. This feature is an outstanding advantage, incidentally, because in high-horsepower machinery, it allows for expansion of the shaft due to thermal effects.

Supports   One of our first problems in this chapter is to devise simpler methods of representing the various methods by which beams are supported. Actually, this is not very difficult because there are only three basic ways of supporting them. These are:

1   The *roller* support
2   The *hinged* or *pinned* support
3   The *fixed* or *built-in* support

Figure 4-2a shows a beam supported by two columns. There is a pin support at the left end and a roller support at the right. We might represent this beam diagrammatically as in Fig. 4-2b. Note that the worm shaft of Fig. 4-1c is supported in exactly the same manner. Since the shaft is supported by ball bearings,

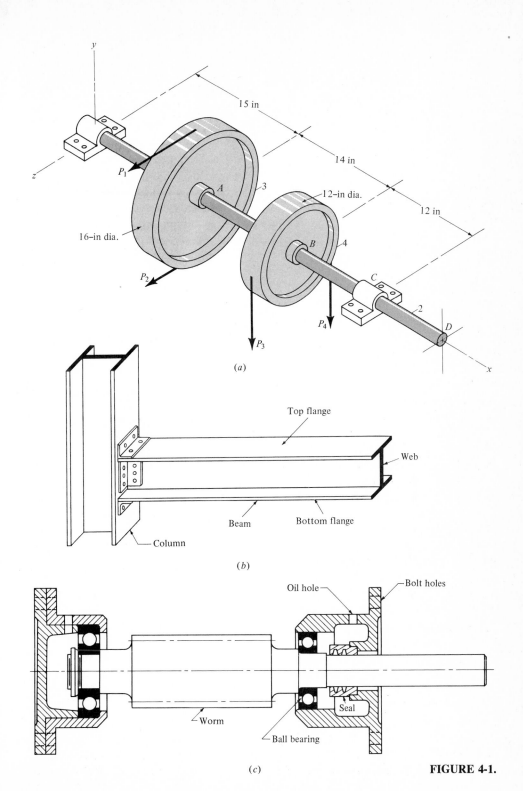

y

z

15 in

14 in

12-in dia.

12 in

P_1

A

3

16-in dia.

B

4

P_2

C

2

D

P_4

x

P_3

(*a*)

Top flange

Web

Beam

Bottom flange

Column

(*b*)

Oil hole

Bolt holes

Worm

Seal

Ball bearing

(*c*)

FIGURE 4-1.

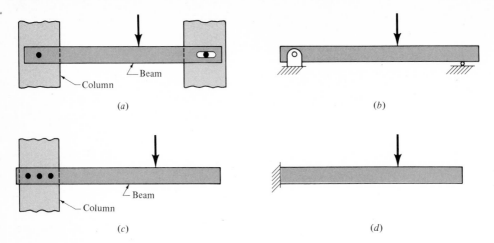

FIGURE 4-2.

the bearings are self-aligning. The right-hand bearing floats and so it acts exactly the same as a roller support. Note that a roller support can only transfer a transverse load into the frame. The left-hand bearing of Fig. 4-1c transfers both thrust and transverse loads into the frame and so serves the same purpose as a hinge.

To summarize, a roller support can take only transverse loads. A hinge support can take transverse loads and axial loads. The hinge support and the roller support are called *simple supports,* and the beam of Fig. 4-2b is said to be a *simply supported beam.*

Figure 4-2c shows a beam supported by a single column. This is called a *fixed* or a *built-in support.* Though three rivets are shown, it could be considered as fixed if only two rivets had been used. A fixed support can transfer two forces, transverse and axial, and a moment into the frame. This beam is called a *cantilever;* it is said to be *cantilevered* to the column. Figure 4-2d is a simplified representation of a beam fixed at one end.

The connection of the beam to the column in Fig. 4-1b is called a *moment connection* by structural engineers because the connection can transfer both moment and force into the column. Therefore, this beam would be treated as having fixed ends. In case the angle that connects the top flange to the column is omitted, the connection would most probably be treated as a simple support, even though some fixity may still remain.

In most cases, the shaft of Fig. 4-1a would be treated as a beam on simple supports. Such an assumption would lead to more conservative results in design. Depending upon the length, the bearings do offer some possibility of supporting a moment, but wear on the outer edges would eventually decrease the degree of fixity. Thus, it would be best to treat the shaft as having simple supports in the first place.

Loading   In Chap. 1 (Sec. 1-9) we learned that in nature forces usually exist as force fields (see Fig. 1-10). In many, but not all, practical situations we reduce these to force distributions or even to force vectors. These idealizations make

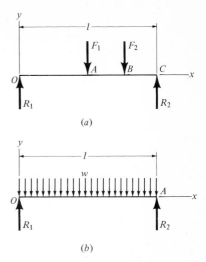

FIGURE 4-3. (a) Simply supported beam with two concentrated loads; (b) simply supported beam with a distributed load.

the problem much easier to solve and generally cause the results to be more conservative than if the idealizations had not been made.

Figure 4-3 shows two simply supported beams. In part a of the figure, the bending is produced by the two concentrated forces or force vectors F_1 and F_2. Note that the effect of the simple supports has been replaced by the support reaction forces R_1 and R_2. A distributed bending load of magnitude w units of force per unit length is shown acting downward in Fig. 4-3b. The support reactions are R_1 and R_2 as before. Since w is a force distributed uniformly from O to A, the total force acting on the beam is

$$F = wl$$

Reference axes   Generally, we shall show beams in a horizontal position as in Figs. 4-2 and 4-3, even though they might occupy a different position in the machine or structure from which they were extracted. Also, we shall usually employ the coordinate system shown for the beams of Fig. 4-3. Here x designates the distance from the origin O to any point along the beam. The use of the coordinate y will be described later, but the direction of y indicates the positive direction for forces. Thus, R_1 and R_2 are considered to be positive; and F_1 and F_2 are negative forces because they act in the direction of negative y. The distributed load w is also shown acting in the negative direction. A particular value for w might be

$$w = -120 \text{ lb/in}$$

Kinds of beams   We have already learned that a *cantilever* is a beam with one end fixed and the other end free. And that a simply supported beam is one that has two simple supports. If there are more than two supports, then the beam is called a *continuous beam* and the distance between supports is called the *span*. *Curved beams,* such as *C* frames and torsion helical springs, will not be treated in this book.[1] A *statically determinate beam* is one in which all the reactions can be found by using the principles of static equilibrium. But sometimes there are some unknown reactions after we have used all of the statical relations available to us. Such beams are called *statically indeterminate.* We shall consider these in Chap. 8.

4-2. FORCE ANALYSIS OF BEAMS WITH SIMPLE SUPPORTS   While this section is merely a review of the material already covered in Chap. 3 (see Secs. 3-1 and 3-2), there is a possibility of developing a more organized approach to the problem. Figure 4-4 shows the general case of a simply supported beam with known forces F_1, F_2, and F_3 and unknown reactions R_1 and R_2. If we were to write Eq. (3-4) for the sum of forces as

$$\Sigma F_y = R_1 - F_1 - F_2 + R_2 - F_3 = 0 \qquad (a)$$

we could not solve it because R_1 and R_2 are both unknown. By using Eq. (3-5) for the sum of moments, it is possible to select the moment axis so that one of the unknowns has a zero moment. Thus, if, in Fig. 4-4, we select a moment axis at O, we have

$$\Sigma M_O = -aF_1 - (a + b)F_2 + lR_2 - (l + d)F_3 = 0 \qquad (b)$$

Since all dimensions are known, the only unknown appearing in Eq. (*b*) is R_2, and so the equation can be solved. With R_2 known, we can solve Eq. (*a*) next for R_1.

Note the sign convention used in writing Eq. (*a*). The forces F_1, F_2, and F_3 are acting down, in the negative y direction, and are treated as negative. The reactions R_1 and R_2 were assumed to be acting up and hence were treated as positive forces. If it turns out that Eq. (*a*) or (*b*) yields a negative value for R_1 or R_2, then the assumed direction was incorrect. This happens quite often.

In writing Eq. (*b*), the sign convention used is that clockwise moments are treated as negative, and counterclockwise moments are positive. Thus, the axis of moments could be selected through point C instead. The moment equation would then become

$$\Sigma M_C = -lR_1 + (b + c)F_1 + cF_2 - dF_3 = 0$$

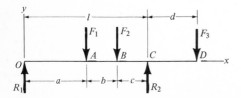

FIGURE 4-4.

[1] See Joseph E. Shigley, "Mechanical Engineering Design," 2d ed., pp. 88–93, McGraw-Hill Book Company, New York, 1972.

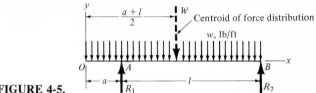

FIGURE 4-5.

This equation can be solved directly for R_1, which is the only unknown. With R_1 known, R_2 can be obtained from Eq. (a).

Figure 4-5 shows a simply supported beam, with an overhanging end, loaded with a uniformly distributed load. The *centroid* of the load distribution (see Appendix A-7) is located at the center of the entire beam length, as shown, because the distribution is rectangular. In making the static force analysis, it is perfectly correct to replace the force distribution with the total force acting at the centroid of the distribution. The magnitude of this force is the same as the area of the force distribution, and is

$$W = w(l + a) \tag{c}$$

If we now select a moment axis through A, Eq. (3-5) becomes

$$\Sigma M_A = -\left(\frac{a+l}{2} - a\right)W + lR_2 = 0 \tag{d}$$

This equation can be solved for R_2 because W is known from Eq. (c). Then for the summation of forces in the y direction, we have

$$\Sigma F_y = R_1 - W + R_2 = 0 \tag{e}$$

When Eq. (e) is solved for R_1, the static force analysis is complete.

EXAMPLE 4-1   Find the reactions R_1 and R_2 for the beam of Fig. 4-6.

SOLUTION

Taking a summation of moments about O gives

$$\Sigma M_O = -20(40) - 30(80) + 40R_2 - 56(120) = 0$$

Therefore,

$$R_2 = \frac{1}{40} [20(40) + 30(80) + 56(120)] = 248 \text{ lb} \qquad Ans.$$

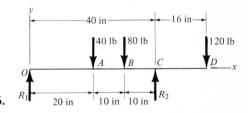

FIGURE 4-6.

Next, we take a summation of forces in the y direction to get

$$\Sigma F_y = +R_1 - 40 - 80 + 248 - 120 = 0$$

Or,

$$R_1 = 40 + 80 - 248 + 120 = -8 \text{ lb} \quad Ans.$$

The negative sign shows that the reaction R_1 acts in the downward direction instead of the direction shown in Fig. 4-6. This negative sign for R_1 has nothing to do with the sign convention for forces. In fact, had we shown the reaction R_1 as acting *down* in Fig. 4-6, then we would have obtained $R_1 = +8$ lb, indicating that we had assumed the correct direction in sketching the free-body diagram.   ////

EXAMPLE 4-2   Find the reactions R_1 and R_2 for the beam of Fig. 4-7.

SOLUTION

The total load on the beam is

$$W = wl = 150(3 + 12) = 2250 \text{ lb}$$

The centroid of the load distribution is at the center of the distribution. This is located at

$$x = \frac{3 + 12}{2} = 7.5 \text{ ft}$$

from the origin O. Taking a summation of moments about A gives

$$\Sigma M_A = -(7.5 - 3)(2250) + 12R_2 = 0$$

and so

$$R_2 = \frac{(7.5 - 3)(2250)}{12} = 844 \text{ lb} \quad Ans.$$

Then when we sum the forces in the y direction to zero, we get for R_1

$$\Sigma F_y = R_1 - 2250 + 844 = 0$$
$$R_1 = 2250 - 844 = 1406 \text{ lb} \quad Ans. \qquad ////$$

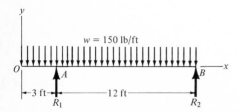

$w = 150$ lb/ft

FIGURE 4-7.

4-3. INCREMENTS, LIMITS, AND RATES   Any quantity for which it is possible to assign an unlimited number of values is called a *variable*. In the equation of a straight line

$$y = mx + b \tag{a}$$

y and x are variables because they represent the coordinates of any point on the line. The factors m and b in the equation are *constants* of the line. A *constant* is a quantity having a fixed value.

In dealing with various quantities, the terms *independent variable* and *dependent variable* occur. In Eq. (*a*), if we say that y, the first variable, is a function of the second variable x, then this implies that we are free to choose any value we please for x and that the value of y then *depends* upon x. In this context, x is the *independent variable* and y is the *dependent variable*.

By saying that y is a function of x, we mean that the value of y is dependent upon the particular value selected for x. Thus, tensile stress is a function of the tensile force applied to a bar because the stress depends upon the value of the force. Note that the stress is also a function of the cross-sectional area; so stress is the dependent variable and area, or force, the independent variable.

It is customary to use the symbol $f(x)$ to designate a function of x. It is read f of x. For example, if

$$y = x^2 + 3x \tag{b}$$

then it is appropriate to say that y is a function of x and write

$$y = f(x) = x^2 + 3x \tag{c}$$

If we wish to know the value of $f(x)$ for a particular value of x, say $x = 4$, then we write

$$f(4) = (4)^2 + 3(4) = 28$$

Similarly, if $x = a + 2$, then $f(x)$ for this value of x is

$$\begin{aligned} f(a + 2) &= (a + 2)^2 + 3(a + 2) \\ &= (a^2 + 4a + 4) + (3a + 6) \\ &= a^2 + 7a + 10 \end{aligned}$$

When we change a variable from one numerical value to another, the difference between the two numerical values is called the *increment*. If x_1 is the first value and x_2 the second, then the increment of x is written as

$$\Delta x = x_2 - x_1 \tag{d}$$

where the symbol Δx is read *delta x*. Do not read this symbol as *delta times x*, because it doesn't have this meaning at all. It simply means or stands for the distance measured along the x axis from x_1 to x_2. This increment can have either positive or negative values, depending upon whether x is getting larger or smaller.

If, in the function $y = f(x)$, x is increased by the increment Δx, then the symbol Δy means the corresponding increment of y. Thus, if

$$y = 2x^2 + 3x$$

and x increases from $x_1 = 4$ to $x_2 = 7$, then

$$\Delta x = x_2 - x_1 = 7 - 4 = 3$$

and $\Delta y = y_2 - y_1$
$$\begin{aligned}
&= (2x_2{}^2 + 3x_2) - (2x_1{}^2 + 3x_1) \\
&= [2(7)^2 + 3(7)] - [2(4)^2 + 3(4)] \\
&= 119 - 44 = 75
\end{aligned}$$

This means that if x takes on an increment $\Delta x = 3$ when $x = x_1$, then the corresponding change in y is $\Delta y = 75$.

Often we don't want to specify a particular value for x_1, but instead wish to choose an increment Δx beginning from an initial point x, instead of x_1, it being understood that a fixed initial value will eventually be specified for x. Under these conditions, y will take on a corresponding increment Δy. As an example, suppose we have given the function

$$y = 2x^2 \qquad\qquad (e)$$

Then $\Delta y = y_2 - y$
$$= 2x_2{}^2 - 2x^2 \qquad\qquad (f)$$

But $x_2 = x + \Delta x$

Therefore, Eq. (f) becomes

$$\begin{aligned}
\Delta y &= 2(x + \Delta x)^2 - 2x^2 \\
&= 2[x^2 + 2x\Delta x + (\Delta x)^2] - 2x^2 \\
&= 2x^2 + 4x\Delta x + 2(\Delta x)^2 - 2x^2 \\
&= 4x\Delta x + 2(\Delta x)^2 \qquad\qquad (g)
\end{aligned}$$

Equation (g) makes it possible to compute Δy for any initial value of x and increment Δx. For example, if $x = 5$ and $\Delta x = 2$, we have

$$\begin{aligned}
\triangle y &= 4x\Delta x + 2(\Delta x)^2 \\
&= 4(5)(2) + 2(2)^2 \\
&= 48
\end{aligned}$$

But if we take the increment $\Delta x = 2$ beginning at $x = 1$, then

$$\Delta y = 4(1)(2) + 2(2)^2 = 16$$

Thus, the change in y is three times as large at $x = 5$ as it is at $x = 1$.

Now let us utilize these increments of x and y to find the slope of a graph. In Fig. 4-8, the function $y = 2x^2$ has been plotted for values of x between 0 and 5. At $x = 0$, the slope of the graph appears to be horizontal. As we progress from $x = 0$ to $x = 5$, which locates point A, we see that the graph gets steeper; that is to say, the slope increases as we move in the positive direction from the origin.

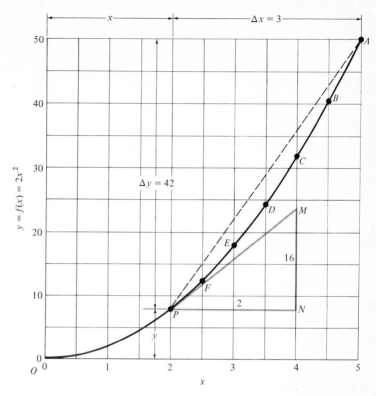

FIGURE 4-8.

If we take an increment of x, $\Delta x = 3$, beginning at P, then the corresponding increment of y is seen from the graph to be $\Delta y = 42$. The *average slope* of the graph between points P and A is

$$\frac{\Delta y}{\Delta x} = \frac{42}{3} = 14$$

We use the word "average" to describe this slope because the slope is less than this at P and more at A.

Rather than measuring increments on a graph, it is often more convenient to develop a formula for the slope. In general we know that

$$\Delta y = y_2 - y \qquad \text{and} \qquad x_2 = x + \Delta x$$

Therefore,

$$\Delta y = f(x_2) - f(x)$$
$$= f(x + \Delta x) - f(x)$$

And so a formula for the average slope of any function $y = f(x)$ is

$$\frac{\Delta y}{\Delta x} = \frac{f(x + \Delta x) - f(x)}{\Delta x} \tag{4-1}$$

TABLE 4-1

line segment	x	Δx	$\dfrac{\Delta y}{\Delta x}$
PA	2	3	14
PB	2	2.5	13
PC	2	2	12
PD	2	1.5	11
PE	2	1	10
PF	2	0.5	9

Substituting the function (of Fig. 4-8) $f(x) = 2x^2$ into Eq. (4-1) yields

$$\frac{\Delta y}{\Delta x} = \frac{2(x + \Delta x)^2 - 2x^2}{\Delta x}$$
$$= \frac{2[x^2 + 2x\Delta x + (\Delta x)^2] - 2x^2}{\Delta x}$$
$$= \frac{2x^2 + 4x\Delta x + 2(\Delta x)^2 - 2x^2}{\Delta x}$$
$$= 4x + 2\Delta x \tag{h}$$

Using this formula, instead of the graph, we find the average slope between points P and A to be

$$\frac{\Delta y}{\Delta x} = 4(2) + 2(3) = 14$$

which checks with our previous result. It will be interesting to use Eq. (*h*) to compute the slopes between point P and the other points shown on the graph. When we do this, we get the results shown in Table 4-1. This tabulation shows how the average slope decreases as we take shorter and shorter line segments.

The instantaneous slope of the curve at P can be obtained by drawing the tangent PM as shown in Fig. 4-8. The slope of this tangent is the instantaneous slope of the curve at P and is seen to be

$$\frac{MN}{PN} = \frac{16}{2} = 8$$

It is also apparent from Table 4-1 that, as Δx becomes smaller, the average slope is getting close to 8. We therefore wonder what would happen if we permitted Δx to get smaller and smaller. Of course, we cannot let $\Delta x = 0$ because then Eq. (4-1) would be indeterminate. We therefore say that we want Δx to *approach zero*. This is usually written $\Delta x \to 0$ and means that Δx is to approach as closely to zero as we please. Under these conditions, Eq. (4-1) would be written

$$\operatorname*{Limit}_{\Delta x \to 0} \frac{\Delta y}{\Delta x} = \operatorname*{Limit}_{\Delta x \to 0} \frac{f(x + \Delta x) - f(x)}{\Delta x} \tag{4-2}$$

This expression means to find the *limiting value* of $\Delta y/\Delta x$ by letting Δx approach zero. The result gives the *instantaneous slope* of the function $f(x)$ at x.

If we apply this technique to Eq. (h), we get

$$\underset{\Delta x \to 0}{\text{Limit}} \frac{\Delta y}{\Delta x} = \underset{\Delta x \to 0}{\text{Limit}} \ (4x + 2\Delta x) = 4x$$

Thus, the instantaneous slope of the graph at P is 8. Using the same formula, we find the instantaneous slope at A to be 20.

It is also appropriate to describe the slope of the graph $y = f(x)$ as the *rate of change of y with respect to x*. Thus, in Fig. 4-8, the rate of change of y with respect to x at point P is 8.

4-4. DIFFERENTIATION   The instantaneous rate of change of a variable y with respect to another variable x is called the *derivative of y with respect to x*. This derivative is usually designated as $f'(x)$, or as dy/dx, and frequently in a variety of other ways. Using Eq. (4-2), we see that the derivative is the limit

$$f'(x) = \frac{dy}{dx} = \underset{\Delta x \to 0}{\text{Limit}} \frac{\Delta y}{\Delta x} = \underset{\Delta x \to 0}{\text{Limit}} \frac{f(x + \Delta x) - f(x)}{\Delta x} \qquad (4\text{-}3)$$

The symbol dy/dx is read *dy over dx*. The terms dy and dx are called *differentials*. Thus, dy is the *differential of y* and dx is the *differential of x*. Using Eq. (4-3), we could write that

$$dy = f'(x) \ dx \qquad (4\text{-}4)$$

which only states that the product of the derivative of y with respect to x, times differential x, is equal to differential y. Mostly we think of the differentials dy and dx in the same manner as the increments Δy and Δx.

Table A-5 lists worked-out differentials for a number of common functions. The examples that follow illustrate how these differentials were obtained from Eq. (4-3) and how the table may be used to find other derivatives.

EXAMPLE 4-2   Differentiate $y = 4x$ with respect to x.

SOLUTION

$$\Delta y = 4(x + \Delta x) - 4x$$
$$= 4 \ \Delta x$$
$$\frac{\Delta y}{\Delta x} = 4$$
$$\frac{dy}{dx} = \underset{\Delta x \to 0}{\text{Limit}} \frac{\Delta y}{\Delta x} = 4 \qquad Ans. \qquad ////$$

EXAMPLE 4-3   Find the derivative of y with respect to x if $y = x^2$.

SOLUTION

$$\Delta y = (x + \Delta x)^2 - x^2$$
$$= x^2 + 2x(\Delta x) + (\Delta x)^2 - x^2$$
$$= 2x(\Delta x) + (\Delta x)^2$$

$$\frac{\Delta y}{\Delta x} = 2x + \Delta x$$

$$\frac{dy}{dx} = \lim_{\Delta x \to 0} \frac{\Delta y}{\Delta x} = 2x \qquad Ans.$$

This means that $dy = 2x\,dx$ (see Table A-5).   ////

EXAMPLE 4-4   What is the slope of the curve of y with respect to x if $y = 1/x^2$?

SOLUTION

$$\Delta y = f(x + \Delta x) - f(x)$$

$$= \frac{1}{(x + \Delta x)^2} - \frac{1}{x^2}$$

$$= \frac{x^2 - (x + \Delta x)^2}{(x + \Delta x)^2 x^2}$$

$$= -\frac{\Delta x(2x + \Delta x)}{(x + \Delta x)^2 x^2}$$

$$\frac{\Delta y}{\Delta x} = -\frac{2x + \Delta x}{(x + \Delta x)^2 x^2}$$

$$\frac{dy}{dx} = \lim_{\Delta x \to 0} \frac{\Delta y}{\Delta x} = -\frac{2x}{x^2 x^2} = -\frac{2}{x^3} \qquad Ans.$$

This result can be expressed in differential form as

$$d\left(\frac{1}{x^2}\right) = -\frac{2}{x^3}\,dx \qquad\qquad ////$$

EXAMPLE 4-5   Use Table A-5 to find the derivative of $y = 1/x^4$.

SOLUTION

From Table A-5, we find the formula

$$d(x^n) = nx^{n-1}\,dx$$

Therefore,

$$\frac{d}{dx}(x^n) = nx^{n-1}$$

In this case $n = -4$, and so

$$\frac{d}{dx}\left(\frac{1}{x^4}\right) = -4x^{(-4-1)} = -\frac{4}{x^5} \qquad Ans.$$

////

EXAMPLE 4-6   Use Table A-5 to differentiate the function

$$y = 4 - \frac{2}{x} + 3x^3$$

SOLUTION

We can write this equation in the form

$$y = 4x^0 - 2x^{-1} + 3x^3$$

and use the formula

$$d(x^n) = nx^{n-1}\, dx$$

for each term. This gives

$$\frac{dy}{dx} = 4(0x^{-1}) - 2(-1x^{-1-1}) + 3(3x^{3-1})$$

$$= 0 + 2x^{-2} + 9x^2$$

$$= \frac{2}{x^2} + 9x^2 \qquad Ans.$$

////

4-5. GRAPHICAL DIFFERENTIATION   Sometimes the functions with which engineers must deal are not readily expressed in algebraic or mathematical form. For example, the function may have been obtained from an oscilloscope plot of an experiment. In such cases, differentiation cannot be accomplished by means of Table A-5 and we must resort to numerical or graphical means.

Figure 4-9a is a plot of a function $y = f(x)$. In order to differentiate this function, we must note the scales to which x and y are plotted. These are

$$S_x = \frac{x_{\text{max}}}{l} \qquad S_y = \frac{y_{\text{max}}}{h} \tag{4-5}$$

where x_{max} and y_{max} are the maximum values of x and y, respectively, and l and h are the respective distances measured on the drawing, as shown in the figure. Note that the units of these scales will be the units of the variables divided by the units of the drawing. If, for example, y is in pounds and h is in inches, then s_y is in lb/in. If SI is used, y might be in kilonewtons and h in millimetres. Then S_y would have units of kN/mm.

To differentiate graphically, we are going to obtain an approximate solution to the equation

$$y' = \underset{\Delta x \to 0}{\text{Limit}} \frac{\Delta y}{\Delta x}$$

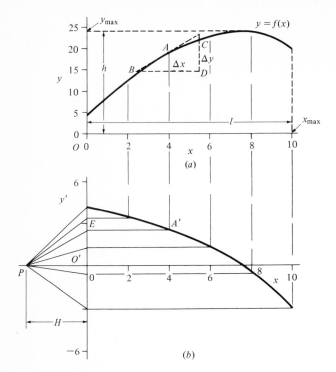

FIGURE 4-9. (*a*) **The function; (*b*) the derivative.**

To accomplish this, select some point, say point A, on the curve of Fig. 4-9*a* and draw the tangent BC. Then the slope of the tangent is the ratio of the ordinate DC to the abscissa BD. Now if we draw additional tangents at other points and construct triangles in a similar manner, and if, in addition, we construct all the triangles so that their abscissas are all the same and equal to BD, then all we must do to obtain the derived curve is to plot the ordinate of each of these triangles. Then, the curve passing through these ordinate values will be the derivative of the original function.

Rather than construct these triangles on the given curve itself, it is expedient to use the same abscissa for all of them and to construct the triangles adjacent to the axes of the derived curve. Thus, in Fig. 4-9*b*, triangle BCD of the function is represented by triangle PEO'. So the distance PO' is the common abscissa for all the triangles. The lines or rays, like PE, are drawn parallel to the various tangents. The distance PO' is called the *pole distance* and its length is designated as H. The point A', located horizontally from E, represents the slope and hence the derivative of the original function at A.

With this method of construction, the pole distance is first selected. It can be of any convenient length. Then tangents are drawn at various points on the given curve and, with instruments, lines are drawn through the pole (point P) parallel to the respective tangents. The intersection of these rays with the y' axis gives the various derivatives.

The relationship between the scales and pole distance is as follows:

$$S_y = HS_x S_{y'} \tag{4-6}$$

where $S_{y'}$ is the scale of the derivative and H is the pole distance as measured on the drawing.

The accuracy of the result depends upon the size of the drawing and upon how well the tangents are constructed. If the curve is very steep, either use more points or increase the length of the horizontal axis to flatten it out.

A very important value of graphical differentiation is its use in estimating and sketching of approximate derivatives. This is particularly useful in predicting slopes and changes of direction in a qualitative, rather than quantitative, manner.

4-6. INTEGRATION[1]   Many inverse operations occur in mathematics. Addition is the inverse of subtraction; division is the inverse of multiplication. In calculus, the *inverse of differentiation is called integration*. Thus, if we know the derivative of a function, then integration is the process of finding the function itself. This function is called the *integral*. Thus,

$$\int f'(x) \, dx = f(x) \tag{a}$$

Here $f(x)$ is the integral of the function $f'(x)$. The differential dx means that the *integration variable* is x. The integral sign is an elongated version of the letter S and means that integration is a summation process.

As an example of Eq. (*a*), suppose

$$f(x) = x^4$$

Then $f'(x) = \dfrac{d}{dx}(x^4) = 4x^3$

Substituting in Eq. (*a*), we have

$$\int 4x^3 \, dx = x^4 \tag{b}$$

This shows that integration and differentiation are inverse operations.

Next, let $f(x) = x^4 + 3$. Then

$$f'(x) = \frac{d}{dx}(x^4 + 3) = 4x^3$$

Consequently,

$$\int 4x^3 \, dx = x^4 + 3 \tag{c}$$

Comparing Eqs. (*b*) and (*c*), it is evident that the integral must be

$$\int 4x^3 \, dx = x^4 + C$$

[1] See also Appendix A-7.

in general, where C is an arbitrary constant, called the *constant of integration.* Thus, the correct form for Eq. (a) is

$$\int f'(x) \ dx = f(x) + C \qquad\qquad\qquad (d)$$

This result is called an *indefinite integral* because the constant C is unknown. In most problems in solid mechanics, the constant C can be evaluated from the known geometry of the problem. The integration process is usually carried out by employing tables of integrals. See Table A-6.

EXAMPLE 4-7   Using Table A-6, integrate the expression

$$y' = 2x^3 - x^2 + 8x - 5$$

SOLUTION

First we write the equation in the form

$$y = \int (2x^3 - x^2 + 8x - 5) \ dx$$

Then by factoring, we get

$$y = 2 \int x^3 \ dx - \int x^2 \ dx + 8 \int x \ dx - 5 \int dx$$

Using the table of integrals separately for each term gives

$$y = 2 \left(\frac{x^4}{4}\right) - \left(\frac{x^3}{3}\right) + 8 \left(\frac{x^2}{2}\right) - 5(x) + C$$

$$= \frac{x^4}{2} - \frac{x^3}{3} + 4x^2 - 5x + C \qquad Ans. \qquad\qquad ////$$

4-7. FORCE ANALYSIS OF BEAMS WITH FIXED SUPPORTS   Abstract examples of beams with fixed supports are shown in Fig. 4-10. Those in Fig. 4-10a are called *cantilevers;* it isn't necessary to say "cantilever beam" because a cantilever *is* a beam. The beams in Fig. 4-10b have no special names; they are described as *beams fixed at both ends.* In each case, the beams on the left are loaded by concentrated loads and those on the right by distributed loads.

Beam ends are said to be fixed if they are fastened such that the ends remain perfectly horizontal during bending of the beam. This means that the ends cannot tilt or rotate even a little bit. But, see also the discussion accompanying Fig. 4-1b. Most of the time we can get the ends fixed by using two or more bolts or rivets (the more the better) and by welding. In many cases, engineers will go to great lengths in design to assure good fixity of the ends. Figure 4-1b is a good example of this.

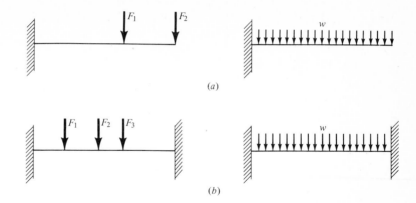

FIGURE 4-10. Examples of beams with fixed supports.

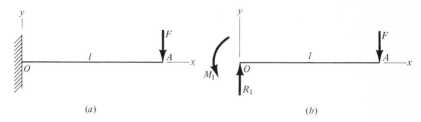

FIGURE 4-11. Force analysis of a cantilever.

Let us make a force analysis now of the cantilever of Fig. 4-11*a*. The free-body diagram is shown in Fig. 4-11*b*. Taking a summation of forces in the vertical direction, we have

$$\Sigma F_y = R_1 - F = 0$$

Therefore, $R_1 = F$ (*a*)

Next, taking a summation of moments about O gives

$$\Sigma M_O = M_1 - Fl = 0$$

and so

$$M_1 = Fl \qquad\qquad\qquad\qquad\qquad\qquad\qquad\qquad (b)$$

Note that F and R_1 are equal, opposite, parallel, and separated by the distance l. Therefore, they constitute a clockwise couple. Then observe that this couple of magnitude Fl is opposed by the counterclockwise moment reaction M_1.

The beams of Fig. 4-10*b* have force reactions and moment reactions at each end. But, we cannot solve for four such reactions using only the principles of statics. Thus, these beams are statically indeterminate. For solutions of these problems, see Chap. 8.

4-8. THE CONCEPT OF INTERNAL ACTIONS AND REACTIONS   In the construction of free-body diagrams, a member or a subsystem of a system is disconnected and isolated in order to determine the forces exerted on the member by the system. By successively applying this technique to each member of the system, we end up with a complete force analysis of every member of the system and of the system itself. Using only a slight variation of this technique provides us with one of the most important and useful tools of solid mechanics; one that we shall use many times in this book.

The method is easy to describe. After a complete force analysis has been made of a single part, we shall want to know something about the forces interior to the part. That is, what is the force that, say, one particle of material exerts on an adjacent particle of material? In order to learn something about the *internal forces* or *stresses,* we imagine that the member is cut into two parts at the cross section of interest. Then we make a free-body diagram of the part that remains and on the cut section show the forces which the removed section must have exerted. Of course this part will be in static equilibrium after cutting, because it was in static equilibrium before cutting. The forces acting on the cut section are then the internal forces necessary to maintain static equilibrium.

To illustrate this analytical tool using numerical values, we select the beam of Fig. 4-12a in which simple supports are shown at O and E. We first make a force analysis of the entire beam to learn that the reactions are $R_1 = R_2 = 500$ lb, as shown in Fig. 4-12b. Though we could separate the beam in two parts anywhere along its length, we arbitrarily select the middle in this example. So, in Fig. 4-12c, the beam has been divided into two portions, each 7 ft long. Note that all of the forces are shown except the internal forces. The internal forces must occur at section C where the beam is imagined to be cut. These internal forces are designated by a question mark because they are the object of this investigation.

Let us assume, until or unless we discover otherwise, that the internal forces consist of a force and a moment. Designate the force as V and the moment as M, and arbitrarily show both of these in the positive direction when applied to the cut section of the left portion. These are shown in Fig. 4-12d. Note that V is directed upward (positive) and M is counterclockwise (positive). These constitute the internal effects of the right portion of the beam acting against the left portion.

The right-hand portion will have the same force V and the same moment M, but they must be opposite in direction. Remember Newton's law. If V and M are the *internal actions* on the left portion, then they are the *internal reactions* on the right-hand portion. If, in our analysis, either V or M should turn out to be negative, then we would know that we had assumed the wrong direction, and hence that a direction opposite to the one shown would be correct.

Now, we make an analysis of the left-hand portion using the principles of statics. First, take a summation of moments about an axis through C

$$\Sigma M_C = -(500)(7) + (100)(6) + (500)(2) + M = 0$$
$$M = +(500)(7) - (100)(6) - (500)(2) = 1900 \text{ lb} \cdot \text{ft}$$

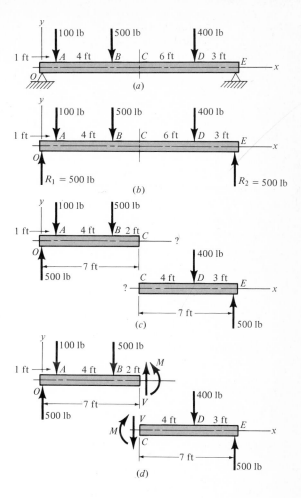

FIGURE 4-12. Free-body diagram of a beam and of portions of the beam.

Next, a summation of forces in the y direction. This gives

$$\Sigma F_y = + 500 - 100 - 500 + V = 0$$
$$V = -500 + 100 + 500 = 100 \text{ lb}$$

Both of our assumptions regarding the directions were correct because V and M are positive. Thus, there is an upward internal force $V = 100$ lb and a ccw internal moment $M = 1900$ lb · ft acting on the cut section of the left half of the beam.

Just to prove that our reasoning is correct, we repeat this analysis for the right-hand portion

$$\Sigma M_C = -M - (400)(4) + (500)(7) = 0$$
$$M = -(400)(4) + (500)(7) = 1900 \text{ lb} \cdot \text{ft}$$
$$\Sigma F_y = -V - 400 + 500 = 0$$
$$V = -400 + 500 = 100 \text{ lb}$$

The results are the same.

Using this technique we could, if desired, investigate the internal reactions from one end of the beam to the other. Such a process would provide us with additional information about the strength or weakness of the beam.

4-9. SHEAR FORCE IN BEAMS   Figure 4-13*a* shows a simply supported beam loaded by a number of external forces *F*. We wish to determine the internal forces on a section of the beam at *C* at some distance *x* from the left support R_1. So we remove the right-hand portion of the beam and replace its effect on the left-hand portion by the reactions *V* and *M* shown in Fig. 4-13*b*. Taking a summation of forces in the *y* direction gives

$$\Sigma F_y = R_1 - F_1 - F_2 + V = 0 \qquad\qquad (a)$$

This equation can also be written in the form

$$\Sigma F_y = \Sigma F_{\text{ext}} + V = 0 \qquad\qquad (4\text{-}7)$$

where $\Sigma F_{\text{ext}} = R_1 - F_1 - F_2$ (b)

The term ΣF_{ext} is called the *external shear force* at section *C* because it consists of the external forces R_1, F_1, and F_2. The term *V* in Eq. (4-7) is called the *internal shear force*. Thus, Eq. (4-7) states that the external shear force plus the internal shear force is zero.

Figure 4-14*a* shows an enlarged portion of the beam on both sides of the section at *C*. For clarity, the moment *M* is omitted. The external shear force ΣF_{ext} pulls the beam upward to the left of the section and pushes it downwards to the right of the section, which explains why ΣF_{ext} is called a *shear force*.

From Eq. (4-7), we see that

$$V = -\Sigma F_{\text{ext}} \qquad\qquad (4\text{-}8)$$

which shows that the internal shear force is equal and opposite to the external shear force. Since we do need a sign convention for shear force, we shall adopt

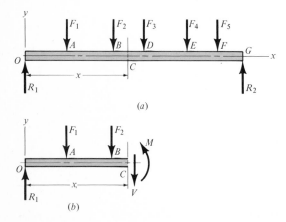

(a)

(b)

FIGURE 4-13.

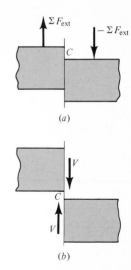

FIGURE 4-14. Definition of positive shear force in beams. (*a*) External shear forces; (*b*) internal shear forces. Both are shown in the positive direction.

the convention shown in Fig. 4-14; all shear forces, both external and internal, are considered as positive. This convention is easy to observe. Simply find the sum of all the forces to the left of the section of interest. This sum is ΣF_{ext}. If this sum acts upward, in the positive y direction, then the shear force is positive. If it acts down, in the negative y direction, then the shear force is negative.

This sign convention requires that the forces be summed to the *left* of the section of interest. The *wrong* sign will be obtained if the forces are summed to the *right* of the section.

4-10. BENDING MOMENT   In summing the moments of the forces acting on a beam about some axis A, we would write, for example,

$$\Sigma M_A = a_1 F_1 + a_2 F_2 + a_3 F_3 + \cdots = 0 \qquad (a)$$

where a_1, a_2, a_3, etc. are the distances from the axis A to the line of action of the forces F_1, F_2, etc. The various terms in Eq. (*a*) are positive for forces that create counterclockwise moments about A and negative for forces that create clockwise moments. The reason for reiterating this convention for problems in statics at this point is that we are about to utilize another, different, convention for the bending of beams.

Figure 4-15*a* shows a simply supported beam loaded by a number of forces F. We imagine the beam to be cut at some section C, a distance x from the origin. Then the effect of the right-hand portion acting on the cut section is an internal shear force V and an *internal resisting moment M,* as shown in Fig. 4-15*b*. The moment M can be obtained by taking moments about any axis. However, if

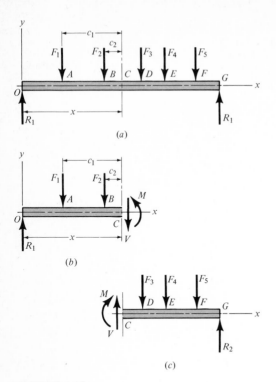

FIGURE 4-15.

moments are taken about an axis through C at the cut section, then the shear force V will not enter into the equation. For Fig. 4-15b, we would then have

$$\Sigma M_C = -xR_1 + c_1F_1 + c_2F_2 + M = 0 \qquad (b)$$

This equation can be written in the form

$$\Sigma M_C = \Sigma M_{\text{ext}} + M = 0 \qquad (4\text{-}9)$$

where $\Sigma M_{\text{ext}} = -xR_1 + c_1F_1 + c_2F_2$

The term ΣM_{ext} is called the *external moment* at section C because it results from the external forces R_1, F_1, and F_2. The term M is the *internal* or *resisting moment*. Equation (4-9) states that the external moment plus the internal moment equals zero. These moments cause the beam to bend, and for this reason are called *bending moments*.

Since the magnitude of the internal moment at a section is the same as the external moment, we will often refer to the moment at a section without specifically distinguishing between them.

The sign convention for bending moment is selected such that both the internal and external moments at section C of Fig. 4-15b and c are treated as positive. Figure 4-16 shows also that a positive bending moment causes the top surface of a beam to be concave. A negative moment causes the top surface to be convex, with one or both ends curved downward.

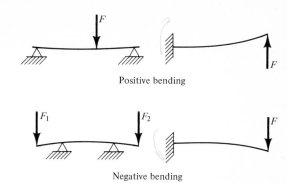

Positive bending

Negative bending

FIGURE 4-16.

As shown in Fig. 4-15*b*, the internal moment acting on the left section is *counterclockwise*. This is a positive bending moment. But note very particularly, from Eq. (4-9), that the external moment ΣM_{ext}, which we have now defined as positive, acts in the *clockwise* direction. This means that an external moment computed for the left-hand section is positive if clockwise. Consequently, an external moment computed for the right-hand section is positive if counterclockwise.

4-11. SHEAR FORCE AND BENDING MOMENT DIAGRAMS   A very useful aid in analyzing a beam is a diagram showing how the shear force at a section varies from one end of the beam to the other. Such diagrams may be drawn freehand or they may be drawn to any convenient scale using instruments. Almost always they are drawn directly below and in line with the free-body diagram of the beam. In this way, it is easy to see the effect of changes in the loading on the shear force. If the diagram is a sketch, then all significant ordinates should be labeled with the magnitude of the force. It is also customary to place a bending moment diagram directly below and in line with the shear force diagram to show how the bending moment varies along the length of the beam. Many of these diagrams can be deduced from the knowledge already available, as we shall demonstrate in the following two examples.

EXAMPLE 4-8   Draw the shear force and bending moment diagrams for the beam of Fig. 4-17 and compute and label all significant values. Neglect the weight of the beam.

SOLUTION

This is a simply supported beam. First we must find the reactions R_1 at O and R_2 at C. Taking moments about O gives

$$\Sigma M_O = -(18)(300) - (24)(200) + 34R_2 = 0$$

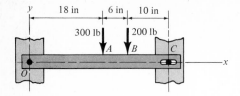

FIGURE 4-17.

Thus,

$$R_2 = \frac{1}{34}\,[(18)(300) + (24)(200)] = 300\text{ lb}$$

Next, take moments about C. We get

$$\Sigma M_C = -34R_1 + (16)(300) + (10)(200) = 0$$

$$R_1 = \frac{1}{34}\,[(16)(300) + (10)(200)] = 200\text{ lb}$$

As a check on these results, write that the summation of forces in the y direction is zero. We get

$$\Sigma F_y = R_1 - 300 - 200 + R_2 = 0$$
$$= 200 - 300 - 200 + 300 = 0$$

which checks.

We begin the diagramming with the loading diagram in Fig. 4-18a. Note that positive forces are in the positive y direction. The shear force diagram is shown in Fig. 4-18b. Review Sec. 4-9 and note that external shear forces acting upward on the left of the section of interest are positive. Choose any section between O and A. The only force acting is $R_1 = 200$ lb, and it acts up positively. So the shear force between O and A is 200 lb, as shown. We are working from the origin O to the right. When we pass point A, we have

$$\Sigma F_{\text{ext}} = +200 - 300 = -100\text{ lb}$$

And so, between A and B, the shear force is -100 lb, as shown. Passing point B, we have

$$\Sigma F_{\text{ext}} = +200 - 300 - 200 = -300\text{ lb}$$

which is the shear force between B and C. Finally, when we pass through point C, the upward force $R_2 = 300$ lb brings the shear-force diagram back to zero, as shown.

Coming now to the bending moment diagram (Fig. 4-18c), remember that a clockwise moment on the left portion acting about the cut section is a positive bending moment. Consider a section at $x = 9$ in from O. The bending moment is

$$\Sigma M_{\text{ext}} = (200)(9) = 1800\text{ lb}\cdot\text{in}$$

which is positive. At 18 in from O, the moment is

$$\Sigma M_{\text{ext}} = (200)(18) = 3600\text{ lb}\cdot\text{in}$$

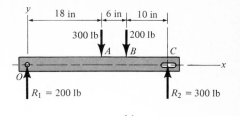

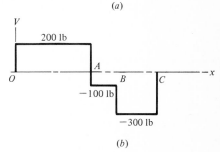

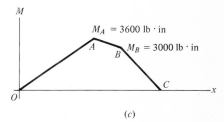

FIGURE 4-18. (*a*) Loading diagram; (*b*) shear-force diagram; (*c*) bending-moment diagram.

This is twice as much as at $x = 9$ in, and so the moment diagram is a straight line between O and A, as shown in Fig. 4-18*c*. At point B, the moment is

$$\Sigma M_{ext} = (200)(18) - (300)(6) = 3000 \text{ lb} \cdot \text{in}$$

and so we draw a straight line from A to B, as shown. Since we know the moment at C is zero, we draw another straight line from B to C to complete the diagram.

$$////$$

EXAMPLE 4-9   Sketch the shear-force and bending moment diagrams of the cantilever of Fig. 4-19. The weight of the beam is included in the distributed load w.

SOLUTION

Figure 4-20*a* shows a free-body diagram of the beam. The beam is bolted to the column with two bolts, and so both moment and shear can be transferred to the column. First, we take moments about the origin. This gives

$$\Sigma M_O = M_1 - wl\left(\frac{l}{2}\right) = 0$$

$$M_1 = \frac{wl^2}{2} = \frac{100}{2}(10)^2 = 5000 \text{ lb} \cdot \text{ft}$$

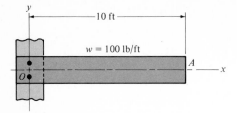

FIGURE 4-19.

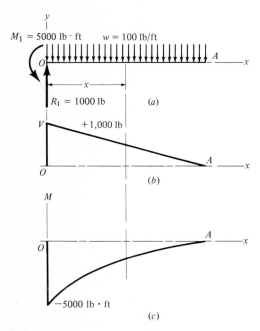

FIGURE 4-20. (*a*) **Loading diagram;** (*b*) **shear-force diagram;** (*c*) **bending-moment diagram.**

Note that M_1 is in the counterclockwise (ccw) direction. The above is an equation of statics in which the rule is that ccw moments are treated as positive. Next, we sum the forces in the y direction

$$\Sigma F_y = R_1 - wl = 0$$
$$R_1 = wl = (100)(10) = 1000 \text{ lb}$$

The shear-force diagram is shown in Fig. 4-20*b*. At a distance x from the origin, the shear force is

$$\Sigma F_{ext} = R_1 - wx = +1000 - 100x$$

Thus, the uniform load w acts in the negative direction and reduces the shear force in direct proportion to the distance x. We know that the shear force is zero at A, and so the shear diagram must be the triangle as shown.

Figure 4-20c shows that the bending moment is negative at the origin (see Sec. 4-10). At a distance x, the bending moment is

$$\Sigma M_{\text{ext}} = -5000 + 1000x - 100x \left(\frac{x}{2}\right)$$
$$= -5000 + 1000x - 50x^2$$

At the midpoint of the beam, $x = 5$ ft and so

$$\Sigma M_{\text{ext}} = -5000 + (1000)(5) - 50(5)^2 = 1250 \text{ lb} \cdot \text{ft}$$

Therefore, the moment diagram is curved, as shown in Fig. 4-20c. ////

4-12. LOADING, SHEAR, AND MOMENT RELATIONS The cantilever illustrated in Fig. 4-21a has a continuously distributed load $q = f(x)$ acting in the positive y direction. We call q the *load intensity* because it is defined by the equation

$$q = \lim_{\Delta x \to 0} \frac{\Delta F}{\Delta x} = \frac{dF}{dx} \qquad (a)$$

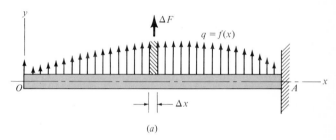

(a)

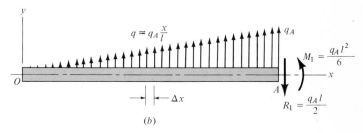

(b)

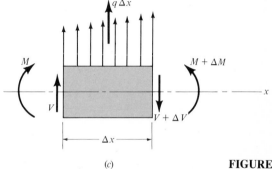

(c)

FIGURE 4-21.

Equation (*a*) shows that the units of *q* are units of force per unit of length. Thus, *q* could be in pounds per inch, pounds per foot, or, say, kilonewtons per metre. In practical engineering situations, such loads arise from gas or fluid pressures, from centrifugal forces, and from gravitational forces.

In the free-body diagram of the cantilever in Fig. 4-21*b*, we have selected a load distribution having a triangular form. The formula for *F* as a function of *x* is

$$F = \frac{q_A x^2}{2l} \tag{b}$$

Then, using Eq. (*a*), we find

$$q = \frac{dF}{dx} = \frac{q_A x}{l} \tag{c}$$

Thus, at $x = 0$, $q = 0$, and at $x = l$, $q = q_A$.

In order to develop general relationships between the loading *q* and the shear and moment, we cut a section of the beam of width Δx and magnify its size in order to find all the forces and moments which act. This section is shown in Fig. 4-21*c* and is called a *differential element*. The left face of the element has a positive moment *M* and a positive shear force *V* acting upon it. The bending force acting on the element is the product of the load intensity *q* and the length of the section and is $q \, \Delta x$, as shown. Because of this bending force, the shear force and bending moment on the right-hand face of the differential section will be increased, and so we designate these as $V + \Delta V$ and $M + \Delta M$, respectively. Taking a summation of the forces in the *y* direction gives

$$\Sigma F_y = V + q \, \Delta x - (V + \Delta V) = 0$$

or $q \, \Delta x = \Delta V$

Solving for *q* gives

$$q = \frac{\Delta V}{\Delta x}$$

If we proceed to the limit, we get the important relation

$$\frac{dV}{dx} = q \tag{4-10}$$

To determine the moment relation, we sum the moments about the right-hand face. Using ccw as the positive direction, we get

$$\Sigma M = -M - V \, \Delta x - (q \, \Delta x)\left(\frac{\Delta x}{2}\right) + (M + \Delta M) = 0$$

Canceling and collecting terms gives

$$\Delta M = V \, \Delta x + q \, \frac{(\Delta x)^2}{2}$$

Dividing both sides by Δx gives

$$\frac{\Delta M}{\Delta x} = V + q\,\frac{\Delta x}{2}$$

Now take the limit of both sides by letting Δx approach zero. The left side becomes

$$\underset{\Delta x \to 0}{\text{Limit}}\ \frac{\Delta M}{\Delta x} = \frac{dM}{dx}$$

And for the right side of the equation, we have

$$\underset{\Delta x \to 0}{\text{Limit}}\ \left(V + q\,\frac{\Delta x}{2}\right) = V$$

Therefore,

$$\frac{dM}{dx} = V \tag{4-11}$$

Equations (4-10) and (4-11) tell us that the loading q is the rate of change with respect to x of the shear force, and that the shear force V is the rate of change of the bending moment.

We might note from our previous work that V changes abruptly when we pass a concentrated force in moving along the length of the beam. Since V has one value to the left of the force and a different value to the right, the derivative dM/dx has a sudden change in magnitude at the location of a concentrated force.

4-13. INTEGRATING THE BEAM EQUATIONS[1]   In this section we shall show how to obtain the shear-force and bending-moment diagrams by integration. First, rearrange Eq. (4-10)

$$dV = q\ dx \tag{a}$$

Next, integrate both sides between any chosen lower limit x_A and any upper limit x_B. Let the shear force be designated as V_A at x_A and V_B at x_B. The left-hand side of Eq. (*a*) then becomes

$$\int_{V_A}^{V_B} dV = V_B - V_A \tag{b}$$

Setting this equal to the integral of the right-hand side gives

$$V_B - V_A = \int_{x_A}^{x_B} q\ dx \tag{4-12}$$

To understand how this equation is used, refer to the generalized loading diagram of Fig. 4-22. At a distance x from the left end of the beam, we have

---

[1] If you are not familiar with the concept of a definite integral, you should study Appendix A-7 carefully before beginning this section.

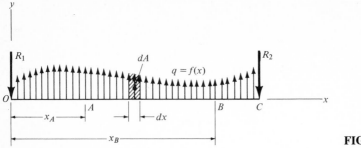

FIGURE 4-22.

selected a differential width dx of the loading diagram. Since the height is q, the differential area is

$$dA = q\,dx \qquad\qquad (c)$$

If we now integrate this expression between A and B, we get for the area

$$A = \int_{x_A}^{x_B} q\,dx \qquad\qquad (d)$$

Equation (d) is identical with the right-hand side of Eq. (4-12) and is the area of the loading diagram between A and B. But the left-hand side of Eq. (4-12) is the difference in the shear force between the two sections A and B of the beam. Thus, Eq. (4-12) states that *the change in shear force from A to B is equal to the area of the loading diagram between x_A and x_B.*

If we now rearrange Eq. (4-11) in a similar manner, we obtain

$$dM = V\,dx \qquad\qquad (e)$$

Integrating the left-hand side between A and B gives

$$\int_{M_A}^{M_B} dM = M_B - M_A \qquad\qquad (f)$$

And, when the right-hand side of Eq. (e) is integrated, we get

$$M_B - M_A = \int_{V_A}^{V_B} V\,dx \qquad\qquad (4\text{-}13)$$

which states that *the change in bending moment from A to B is equal to the area of the shear-force diagram between x_A and x_B.*

EXAMPLE 4-10   Figure 4-23a shows the completed free-body or loading diagram of a beam loaded by three concentrated forces. Such a diagram cannot be integrated to get the shear-force diagram by conventional methods.[1] Consequently, when the forces are concentrated we shall use

---

[1] But it can be integrated using singularity functions; these we shall not introduce in this book. See F. R. Shanley, "Mechanics of Materials," pp. 126–130, McGraw-Hill Book Company, New York, 1967.

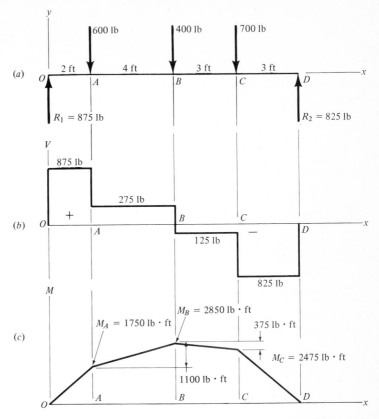

FIGURE 4-23.

the methods of Sec. 4-11 to get the shear-force diagrams. In this example, we want to find the bending-moment diagram by the use of Eq. (4-13).

SOLUTION

The shear-force diagram is easily plotted using the methods already developed. It is shown in Fig. 4-23b. Now we wish to construct the bending-moment diagram using the statement that the change in the bending moment between two points is equal to the area of the shear-force diagram between those same two points. Consider points O and A in Fig. 4-23c. The bending moment at O is zero because the beam rests on simple supports. The shear-force diagram between O and A is 2 ft long and 875 lb high, and so its area is

$$A_{OA} = (2)(875) = 1750 \text{ lb} \cdot \text{ft}$$

Therefore,

$$M_A - M_O = M_A = 1750 \text{ lb} \cdot \text{ft} \qquad Ans.$$

The shear-force diagram between A and B is 4 ft long and 275 lb high. So its area is

$$A_{AB} = (4)(275) = 1100 \text{ lb} \cdot \text{ft}$$

And so

$$M_B - M_A = 1100 \text{ lb} \cdot \text{ft}$$

Therefore, the total bending moment at B is

$$M_B = 1100 + M_A = 1100 + 1750 = 2850 \text{ lb} \cdot \text{ft} \qquad Ans.$$

From B to C, the shear force is negative. The diagram is 3 ft long and 125 lb high between these points. So its area is

$$A_{BC} = -(3)(125) = -375 \text{ lb} \cdot \text{ft}$$

Thus,

$$M_C - M_B = -375 \text{ lb} \cdot \text{ft}$$

And so

$$M_C = -375 + M_B = -375 + 2850 = 2475 \text{ lb} \cdot \text{ft} \qquad Ans.$$

Finally, the shear-force diagram from C to D is 3 ft long and 825 lb high. Its area is

$$A_{CD} = -(3)(825) = -2475 \text{ lb} \cdot \text{ft}$$

Thus,

$$M_D - M_C = -2475 \text{ lb} \cdot \text{ft}$$

And

$$M_D = -2475 + M_C = -2475 + 2475 = 0$$

which checks because the right end of the beam has a simple support too.   ////

4-14. EQUATIONS OF DIAGRAMS   The equations of the diagrams can be obtained by direct integration whenever the function to be integrated is continuous. Roughly speaking, a function $y = f(x)$ is said to be continuous in the interval from $x = a$ to $x = b$ if it is possible to draw a graph in one continuous line without lifting the pencil from the paper. More specifically, a function $y = f(x)$ is continuous at $x = a$ if the limiting value of the function is the value assigned to the function for $x = a$ when x approaches a in *any* manner. Expressed in mathematical terms, a function $f(x)$ is continuous at $x = a$ if

$$\underset{x \to a}{\text{Limit}} f(x) = f(a)$$

Consider the beam of Fig. 4-24a. Here, the loading is

$$q = -w \qquad\qquad\qquad (a)$$

negative because w acts downward. This loading is continuous between O and A. It is discontinuous at O and again at A. For negative values of x, the loading

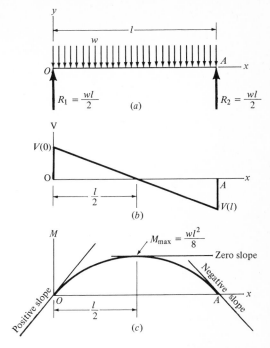

FIGURE 4-24.

is, of course, zero. So exactly at $x = 0$, the loading jumps from $q = 0$ to $q = -w$. A similar situation exists at $x = l$.

If we rearrange Eq. (4-10) as

$$dV = q \, dx$$

and integrate both sides, we have

$$V = \int q \, dx \qquad (4\text{-}14)$$

Note that this is the same as Eq. (4-12) in the preceding section, but without limits of integration. Equation (4-12) is called a *definite integral* because the limits of the integration are specified. Equation (4-14) is called an *indefinite integral*. If we substitute the value of q from Eq. (*a*) and integrate, we get

$$V = \int - w \, dx = -wx + C_1 \qquad (b)$$

where C_1 is the constant of integration. We can find C_1 because we know the shear force at $x = 0$. From Fig. 4-24*b*, we see that the shear force is the same as the support reaction at O. Therefore,

$$V(0) = R_1 = \frac{wl}{2} \qquad (c)$$

If we substitute Eq. (*c*) into (*b*) along with $x = 0$, we get

$$\frac{wl}{2} = -w(0) + C_1$$

or $C_1 = \dfrac{wl}{2}$

Therefore, Eq. (*b*) becomes

$$V = -wx + \frac{wl}{2} \qquad (d)$$

which is the equation of the shear-force diagram between $x = 0$ and $x = l$.

A similar procedure can be used to obtain the equation of the moment diagram. For an indefinite integral, Eq. (4-13) would be written

$$M = \int V \, dx \qquad (e)$$

Upon substituting Eq. (*d*), we obtain

$$M = \int \left(-wx + \frac{wl}{2} \right) dx$$

Integrating by the use of Table A-6 gives

$$M = -\frac{wx^2}{2} + \frac{wlx}{2} + C_2 \qquad (f)$$

Now, we know the bending moment is zero at $x = 0$ because the support at O is a simple one. Substituting $M = 0$ and $x = 0$ in Eq. (*f*) gives $C_2 = 0$. Therefore, the equation for M is

$$M = \frac{w}{2} (lx - x^2) \qquad (g)$$

The maximum moment occurs at the middle of the beam where $x = l/2$. Using Eq. (*g*), we find this to be

$$M_{max} = \frac{w}{2} \left[l \left(\frac{l}{2} \right) - \left(\frac{l}{2} \right)^2 \right] = \frac{wl^2}{8} \qquad (h)$$

Note that the maximum moment occurs where the shear force is zero. Since

$$V = \frac{dM}{dx}$$

then V is the slope of the bending-moment diagram. At $x = l/2$, the shear force is zero and hence the slope of the moment diagram is zero. Therefore, the moment is either a maximum or a minimum because the slope is zero for both; in this case, the moment is maximum.

Note too, that the shear force has its largest positive value at $x = 0$ and, accordingly, the bending-moment diagram has its largest positive slope at $x = 0$.

Also, at $x = l$, the shear force is negative and the bending-moment diagram has a negative slope at this point, as shown in Fig. 4-24c.

4-15. GRAPHICAL INTEGRATION By using instruments and a reasonably large drawing, the various beam diagrams can be obtained with good accuracy by the use of graphical integration. And these same techniques also constitute a useful means of estimating the shapes of the diagrams by free-hand sketching.

Graphical integration can be explained by reference to Fig. 4-25 where the given function $y = f(x)$ is plotted in part a and the integral in part b. As in graphical differentiation, three scales are necessary—one for the dependent variable, one for the independent variable, and one for the integral. These scales are related in the same manner as for differentiation. The scale formula is

$$S_{y_i} = HS_yS_x \tag{4-15}$$

where H = pole distance, in
$\quad S_x$ = scale of x in units of x per inch
$\quad S_y$ = scale of y in units of y per inch
$\quad S_{y_i}$ = scale of integral in units of x times units of y per inch

The integral of a function $y = f(x)$ between $x = a$ and $x = b$ is simply the area below the curve between ordinates erected at $x = a$ and $x = b$. To integrate the function of Fig. 4-25a, choose the line OP, called the *pole distance,* at any convenient length, say 2 in. Project the tops of the various rectangles horizontally to the y axis and draw the rays $P1$, $P2$, $P3$ to each of these intersections. Each ray has a slope proportional to the height of the corresponding rectangle.

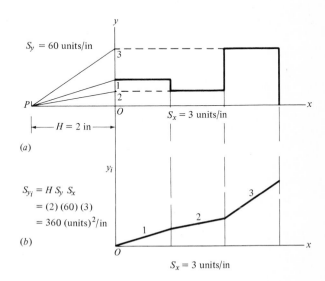

FIGURE 4-25. Graphical integration. The variable x need not be expressed in the same kind of units as is the variable y. (*a*) The function; (*b*) the integral.

The middle rectangle, for example, is short, and the ray $P2$ has only a small slope; but the last rectangle is tall, and so the ray $P3$ is steep.

The integral in Fig. 4-25b is drawn by drawing lines parallel to $P1$, $P2$, and $P3$. Thus, line 1 is parallel to the ray $P1$, line 2 is parallel to $P2$, and line 3 is parallel to $P3$. The scale of the integral is shown in the figure.

When the function to be integrated is a curve, it is divided into a number of vertical strips or segments. Each of these segments is then treated as if it were a rectangle. The integration can then be performed as in Fig. 4-25. Of course, the result is not exact, but the approximation is often quite satisfactory for many practical purposes.

EXAMPLE 4-11 Figure 4-26a is the loading diagram for a beam 20 ft in length on simple supports. The beam is loaded by a uniform load $q = -50$ lb/ft between A and B. This is plotted downward, because it is negative, at a scale of $S_q = 50$ lb/ft per inch of drawing, as shown. The reactions $R_1 = 225$ lb and $R_2 = 275$ lb have already been computed. Integrate the loading diagram to find the shear-force diagram and then integrate the shear-force diagram.

SOLUTION

While the shear-force diagram can be obtained quite easily by the methods already developed in this chapter, we shall use, instead, graphical integration. In Fig. 4-26a, choose $H = 1$ in and compute the scale of the integral using Eq. (4-15). As shown in Fig. 4-26b, this is $S_V = 250$ lb/in of drawing. At $x = 0$, the shear force is the same as the reaction R_1, or 225 lb. So compute

$$\frac{225}{250} = 0.9 \text{ in}$$

and draw line ab 0.9 in above the x axis in Fig. 4-26b. At point b, draw the line bi parallel to the ray $P_1 1$ in Fig. 4-26a. At point i draw the line ij to complete the shear-force diagram. The distance from the x axis to line ij should scale out at 275 lb, and this is a check on the accuracy.

To integrate the shear-force diagram, select a distance, say $H = 1.5$ in. A large value for H will cause the derived diagram to have short ordinates. So the value of H should be chosen such that the derived diagram will have ordinates of a moderate length. The scale of the moment diagram is computed as before and is shown in Fig. 4-26c.

Now, through the shear-force diagram, draw vertical lines through points a, b, c, etc., to j in order to obtain vertical strips that can be treated as rectangles. It is not necessary for these strips to have equal widths. Project the mean ordinate of these strips leftward to locate points 1, 2, 3, etc., to 9. Now, through pole P_2, draw rays $P_2 1$, $P_2 2$, $P_2 3$, etc. Line 1 on the moment diagram is parallel to ray $P_2 1$; line 2 is parallel to $P_2 2$, etc. Line 1 begins at $M = 0$ because the left support is a simple one.

The last line on the moment diagram is line 9, which is parallel to the ray $P_2 9$. It sometimes happens that this line will not "close" the diagram. This is the same as saying that line 9 might happen to intersect the x axis ahead of or after point C. If this should happen, do not attempt to "fudge" on the diagram. Unavoidable inaccuracies due to the graphics will usually result in a failure to close. Figure 4-27 shows what to do when this occurs. Draw a nonhorizontal line from O to the point C' where the moment diagram intersects an ordinate through C. At any point x,

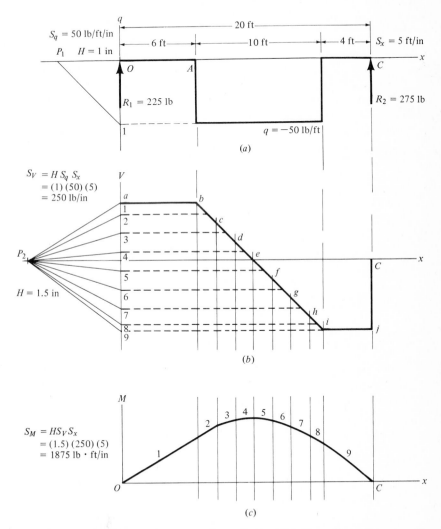

FIGURE 4-26. Shear-force and bending-moment diagrams obtained by graphical integration; $M_{max} = 1856$ lb · ft. (*a*) Loading diagram; (*b*) shear-force diagram; (*c*) bending-moment diagram.

measured parallel to the x axis, the moment M_x is the *vertical* distance from the moment diagram to the line OC', as shown. Figure 4-27 also shows how to find the maximum bending moment. Draw a tangent to the moment diagram parallel to OC'. The tangent point located the point of maximum moment.   ////

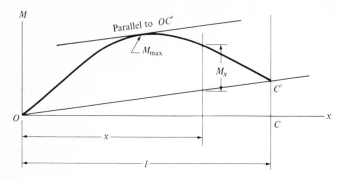

FIGURE 4-27.

PROBLEMS

SECTION 4-2

4-1 to 4-7   Find the magnitude and proper direction for the unknown reactions R_1 and R_2 in each problem. In Prob. 4-2b, find the formulas for R_1 and R_2 in terms of the dimensions a, b, and l, and the applied force F. Note that Probs. 4-4 and 4-7 are in SI units.

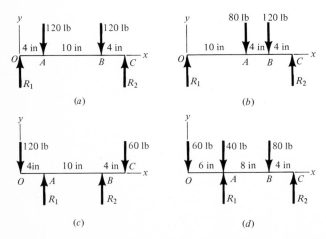

PROB. 4-1.

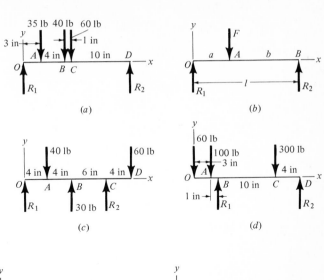

PROB. 4-2.

(a)

(b)

(c)

(d)

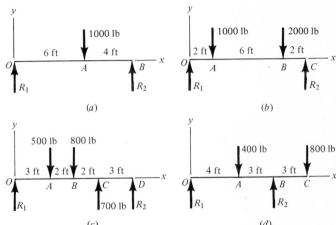

PROB. 4-3.

(a)

(b)

(c)

(d)

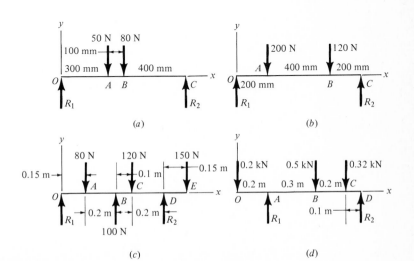

PROB. 4-4.

(a)

(b)

(c)

(d)

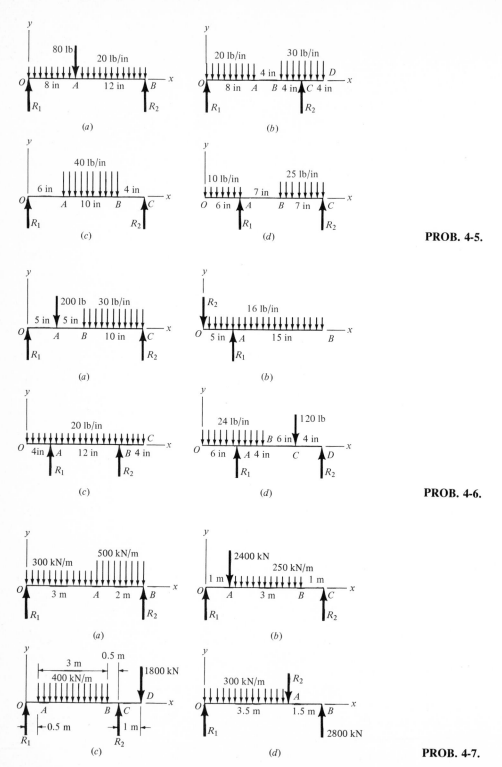

PROB. 4-5.

PROB. 4-6.

PROB. 4-7.

4-8 The figure is a plot of the given equation for x's between 0 and 6. Compute the following using the equations and check your results from the graph:

(a) Find $f(2.8), f(1.5), f(5)$, and $f(6)$.

(b) If $\Delta x = 1.5$, find Δy for $x_1 = 0$, $x_1 = 2.5$, and $x_1 = 4.5$.

(c) Find the average slope of the curve between $x = 0$ and $x = 2$, between $x = 2$ and $x = 4$, between $x = 4$ and $x = 6$.

(d) Find the instantaneous slope of the curve at $x = 1$, $x = 3$, and $x = 5$.

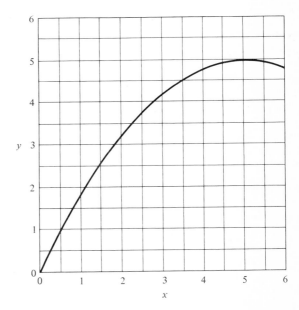

PROB. 4-8. A partial plot of the equation $y = 2x - 0.2x^2$.

4-9 Using the graph and the equation of Prob. 4-8, make the following determinations directly from the graph and check your results using the equation:

(a) Find $f(1), f(2), f(3)$, and $f(4)$.

(b) Using $\Delta x = 1$, find Δy for $x_1 = 0$, $x_1 = 2$, $x_1 = 4$, and $x_1 = 5$.

(c) Find the average rate of change of y with respect to x between $x = 1$ and $x = 2$, between $x = 2$ and $x = 3$, and between $x = 3$ and $x = 4$.

(d) Find the instantaneous rate of change of y with respect to x at $x = 0$, $x = 2$, and $x = 6$.

4-10 Make the following calculations using the equation of Prob. 4-8:

(a) Find $f(-1), f(-2), f(-3)$, and $f(-5)$.

(b) Using $\Delta x = 2$, find Δy for $x_1 = -1$, $x_1 = -3$, and $x_1 = -5$.

(c) Find the average rate of change of y with respect to x between $x = -1$ and $x = +1$, between $x = -2$ and $x = 0$, and between $x = -4$ and $x = -2$.

(d) Find the instantaneous rate of change of y with respect to x at $x = -5$, $x = -3$, and $x = -1$.

4-11   Use the equation

$$y = -4 - 5x + x^2$$

to determine the following:
(a) $f(0), f(-2), f(2)$, and $f(4)$.
(b) Δy when $\Delta x = 1$ and $x_1 = 0$, $x_1 = 2$, and $x_1 = -2$.
(c) The average slope of y between $x = -2$ and $x = 0$, between $x = -1$ and $x = 1$, and between $x = 1$ and $x = 3$.
(d) The instantaneous slope of $f(x)$ at $x = -1$, $x = 0$, $x = 1$, and $x = 4$.

SECTION 4-4
4-12   Using Eq. (4-3), find $f'(x)$ for the following functions: $y = 2/x$, $y = a + bx^2$, $y = x - 3x^2$, $y = 1/x^3$, $y = \pi x^2$.
4-13   Use Table A-5 to solve for the derivatives of Prob. 4-12.
4-14   Use Table A-5 to differentiate the following functions: $y = 2x^3 - 1/\sqrt{x}$, $y = a + bx + cx^2$, $y = a + (b/x) + (c/x^2)$, $y = x\sqrt{x}$.

SECTION 4-7
4-15 to 4-17   Find the support reactions for each beam.

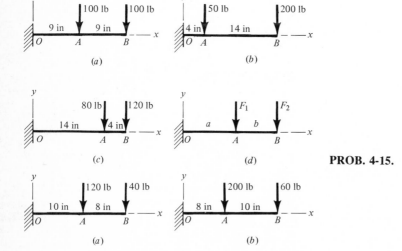

(a)   (b)

(c)   (d)   PROB. 4-15.

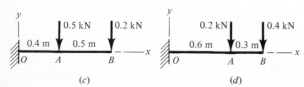

(a)   (b)

(c)   (d)   PROB. 4-16.

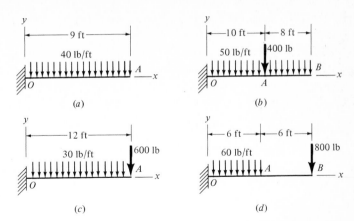

PROB. 4-17.

<small>SECTION</small> 4-11

4-18 to 4-21 Make freehand sketches of the free-body diagram, the shear-force diagram, and the bending-moment diagram for each beam shown in the figure. On these diagrams, record the reactions, and the value of the shear force and bending moment at each change in direction of the respective diagrams.

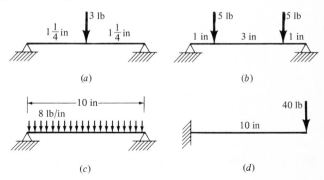

PROB. 4-18.

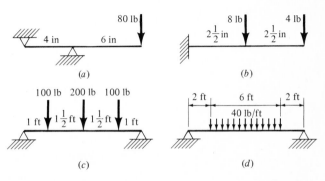

PROB. 4-19.

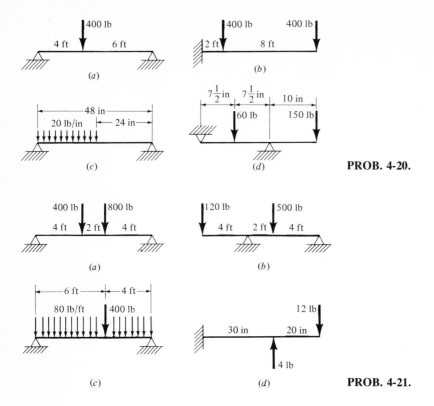

PROB. 4-20.

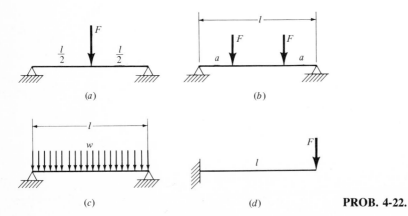

PROB. 4-21.

4-22   Sketch the free-body diagram, the shear-force diagram, and the bending-moment diagram for each beam shown in the figure. Find each reaction in terms of the geometry and loading of the beams. Find the formula for the maximum shear force and maximum bending moment for each beam.

PROB. 4-22.

4-23   Do the same as for Prob. 4-18.

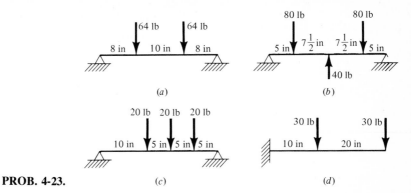

PROB. 4-23.   (a)   (b)   (c)   (d)

SECTIONS 4-14 AND 4-15

4-24 to 4-27

(a) Draw the loading diagram for the beam shown in the figure and find the reactions.

(b) Sketch approximations to the shear-force and bending-moment diagrams by using the techniques of graphical integration.

(c) Find the formulas for the shear-force and bending-moment diagrams wherever these functions are continuous.

(d) What is the maximum bending moment? Where does it occur?

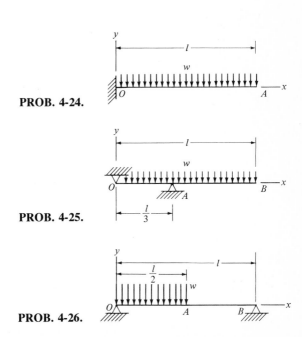

PROB. 4-24.

PROB. 4-25.

PROB. 4-26.

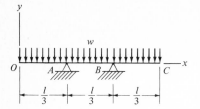

PROB. 4-27.

4-28 Determine the loading diagram with reactions for the beam shown in the figure. Using drawing instruments, plot the loading diagram on a sheet of $8\frac{1}{2}$ by 11 in paper using scales of $S_x = 1$ ft/in and $S_q = 100$ lb/ft/in. Place the origin of the shear-force diagram about $2\frac{1}{2}$ in below the loading diagram. Use $H_q = 1$ in and integrate the loading diagram graphically. Use eight equal-width rectangles on the shear-force diagram, and $H_V = 1$ in. Integrate graphically to obtain the moment diagram and specify the maximum moment and its location.

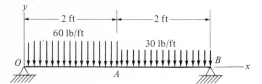

PROB. 4-28.

4-29 Plot the loading diagram for the beam shown in the figure to scales of $S_x = 2$ ft/in and $S_q = 100$ lb/ft/in on $8\frac{1}{2}$ by 11 in paper using instruments. Using graphical integration and $H_q = 1.25$ in, plot the shear-force diagram beginning about 3 in below the loading diagram. Divide the shear-force diagram into 10 equal vertical strips and graphically integrate it using $H_V = 1$ in to get the moment diagram. Start at about 3 in below the shear-force diagram. Record the maximum bending moment, its location, and the bending moment at *A*.

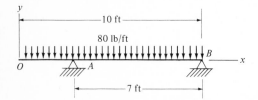

PROB. 4-29.

4-30 The figure shows a simply supported beam that supports a load distribution varying linearly, beginning from zero at the left support to 96 lb/ft at the right support. Find and plot the loading diagram and integrate it successively using graphical techniques to get the shear-force and bending-moment diagrams. Use $S_x = 5$ ft/in, $S_q = 100$ lb/ft/in, and $H_q = H_V = 1$ in. Record the maximum moment and its location. Space the diagrams about $3\frac{1}{2}$ in apart on $8\frac{1}{2}$ by 11 in paper.

PROB. 4-30.

4-31 The figure shows a uniformly loaded, simply supported beam with overhanging ends. Find and plot the loading diagram and graphically integrate twice to get the shear-force and bending-moment diagrams. Use $S_x = 10$ in/in, $S_q = 10$ lb/in/in, and $H_q = H_V = 1$ in. The scale $S_x = 10$ in/in means that 1 in of drawing represents 10 in of beam length. Do not confuse these two kinds of inches. Record the values of the bending moments at both supports and at the middle of the beam.

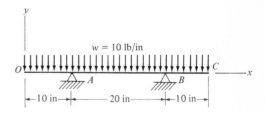

PROB. 4-31.

SI UNITS

4-32 Make freehand sketches of the free-body diagram, the shear-force diagram, and the bending-moment diagram for each beam shown in the figure for Prob. 4-22. Record the reactions and the value of the shear force and bending moment at each change in direction of the respective diagrams. The forces and dimensions are:
  (a) $F = 800$ N, $l = 1.5$ m
  (b) $F = 20$ kN, $l = 4$ m, $a = 0.5$ m
  (c) $w = 250$ N/m, $l = 120$ mm
  (d) $F = 40$ N, $l = 200$ mm

4-33 In the figure for Prob. 4-24, $l = 6$ m and $w = 400$ N/m. Sketch a free-body diagram of the beam and find the reactions at O. Also sketch the shear-force and bending-moment diagrams and find V and M at $x = 0$ and at $x = l$.

4-34 In the figure for Prob. 4-25, $l = 600$ mm and $w = 3000$ N/m. Find the reactions at O and A. Sketch the shear-force and bending-moment diagrams and label completely.

4-35 Use $l = 800$ mm and $w = 2000$ N/m in the figure for Prob. 4-26. Find the reactions at O and B and plot the shear-force and bending-moment diagrams. Compute the maximum bending moment and show where it is located.

4-36 Use $l = 6$ m and $w = 12$ kN/m in the figure for Prob. 4-27 and find the reactions at A and B. Plot the shear-force and bending-moment diagrams and find M_{max}.

bending stresses

5-1 DEFINITIONS AND TERMINOLOGY When a beam is bent, both shear stresses and normal stresses may occur. The shear stresses result from the shear forces. The normal stresses result from the bending moments.

If the internal shear forces are zero in a beam, then the only stresses which result are normal stresses due to the internal bending moment. The beam of Fig. 5-1a is loaded by two forces F, each a distance a from the supports at O and C. This loading yields a shear-force diagram (Fig. 5-1b) with zero shear force between points A and B. As shown in Fig. 5-1c, the bending moment in this section is $M = aF$. The portion of the beam in this section is said to be in *pure bending* because no other effect is present. Note that the forces F and R at each end form a couple at each end having an arm a. These couples can be replaced by moments M at each end, as shown in Fig. 5-1d. In order to eliminate the complicating effects of shear, we shall utilize a beam in pure bending in developing the formula for normal bending stress in this chapter.

In developing the beam theory, we are going to assume that the material is *isotropic* and *homogeneous*. This means that the material must have the same properties in all directions and be of the same nature throughout. Cast iron and concrete are examples of materials that are neither isotropic nor homogeneous. They are not homogeneous because they contain inclusions having different physical characteristics than the matrix. And they are not isotropic because the modulus of elasticity, for example, is different in tension than in compression.

We also require that the beam be *straight*. A special theory is available for the analysis of curved beams, an example of which is a helical torsion spring, but the subject is not included in this book.

Another requirement is that the material must *obey Hooke's law*. This means that the material of the beam must be elastic, have a straight stress-strain line, and that the beam not be bent so much as to result in a stress greater than the elastic limit.

It is also necessary to rule out of our theory beam shapes that might *buckle sidewise* or otherwise *fail locally*. Consider, for example, a beam made of a strip of plywood $\frac{1}{4}$ in thick, 3 in wide, and 8 ft long. If this beam were loaded on the edge, it would buckle sidewise before an appreciable load could be applied in bending.

We must also limit our analysis to beams that have a *constant cross section* throughout the beam length and, at the same time, have an *axis of symmetry in the plane of bending*. Figure 5-2a illustrates some cross sections that satisfy the requirement of symmetry. In all cases, the plane of bending is in the vertical direction; this means that the vertical center line shown is an edge view of the xy plane. Figure 5-2b illustrates some sections that violate the symmetry requirement.

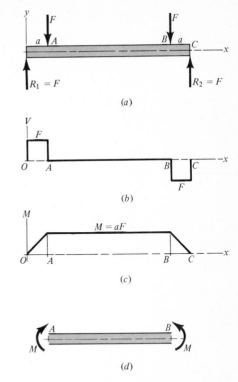

FIGURE 5-1. An illustration of pure bending.

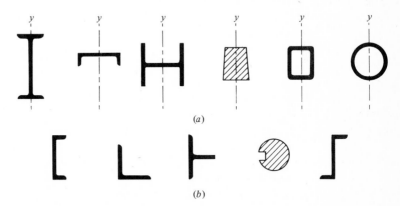

FIGURE 5-2. (*a*) Sections which satisfy the symmetry requirement; (*b*) sections which violate the symmetry requirement. The *y* axis indicates the vertical plane of bending.

Finally, we want to show that plane cross sections in a beam under pure bending remain plane during bending. One can demonstrate this by taking a pencil eraser and drawing straight lines on the edge, as shown in Fig. 5-3*a*. Then, when the eraser is bent (Fig. 5-3*b*), it can be seen that the lines remain straight. There is also another way that we can demonstrate that cross sections remain

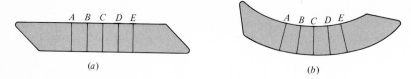

FIGURE 5-3. Straight lines *A, B, C, D,* and *E,* drawn on an eraser, remain straight after bending.

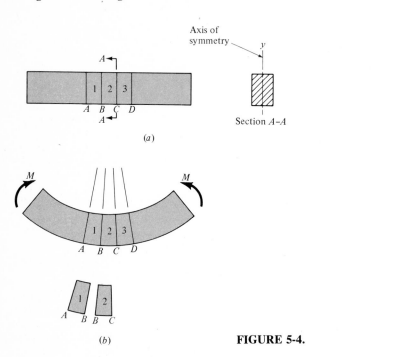

FIGURE 5-4.

plane during bending. Take the beam of Fig. 5-4*a* and divide a part of it into vertical slices 1, 2, and 3 by planes *A, B, C,* and *D*. Now apply a bending moment *M* to the beam causing it to bend as shown in Fig. 5-4*b*. Consider the two adjacent slices 1 and 2. If side *B* on slice 2 is dished in, then side *B* on slice 1 must bulge out, in order to fit. But, if the beam were turned end for end, the bulge would be on the other side. Since the beam is symmetrical end-for-end, we are forced to the conclusion that planes *A, B, C,* and *D* remain plane during bending. This is the only way the end-for-end symmetry requirement can be satisfied.

Figure 5-5 is an enlargement of slice 2 of Fig. 5-4 showing that the plane of symmetry is the *xy* plane. During bending, the top surface of this slice shortens while the bottom surface lengthens. There will be one line, however, which does not change its length. Though we do not yet know the location of this line, we define it as the *neutral axis* and always choose the *x* axis coincident with this line. A plane through the neutral axis perpendicular to the plane of symmetry is called the *neutral plane*. Note, in Fig. 5-5, that the neutral plane is the same as the *xz* plane.

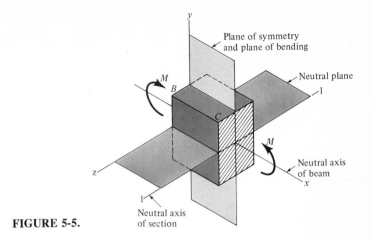

FIGURE 5-5.

Most authors do not distinguish between the terms "neutral axis of beam" and "neutral axis of section." In reading other books, one may find either of these axes called, simply, the neutral axis. Figure 5-5 shows the difference. The *neutral axis of the section* is line 1-1, an edge view of the neutral plane. The neutral axis of the section remains a straight line during bending, but the neutral axis of the beam becomes curved.

5-2. THE FLEXURE FORMULA   In Fig. 5-6*a* is shown a portion of a beam in pure bending subjected to a positive bending moment M. The upper surface of the beam is bent downward and the neutral axis of the beam is curved, as shown. Due to the curvature, a section AB originally parallel to CD, because we are dealing only with beams that are originally *straight,* will rotate through the differential angle $d\phi$ to $A'B'$.[1] Since AB and $A'B'$ are both straight lines, we have utilized the finding that plane sections remain plane during bending. Let us now

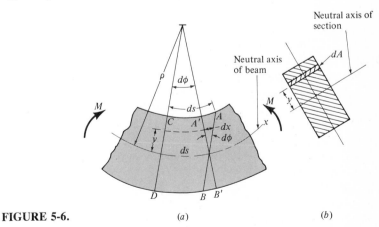

FIGURE 5-6.   (*a*)   (*b*)

[1] See Table A-4 for the names of the Greek letters.

specify the radius of curvature of the neutral axis as ρ and the length of a differential element of the neutral axis of the beam as ds, as shown in Fig. 5-6a. The angle subtended by the two adjacent sides $A'B'$ and CD at the center of curvature O is also $d\phi$. From the definition of radian measure, this angle is obtained from the equation

$$ds = \rho \, d\phi \tag{a}$$

Consider next the compressive deformation of the beam at a distance y from the neutral surface. The deformation of this element is the differential distance dx shown in the figure. Using the definition of radian measure again, we observe that

$$dx = y \, d\phi \tag{b}$$

Since unit strain is equal to the total deformation dx divided by the original length ds, we have

$$\epsilon_x = -\frac{dx}{ds} = -\frac{y \, d\phi}{\rho \, d\phi} = -\frac{y}{\rho} \tag{c}$$

where the negative sign indicates that this is a compressive strain. But Hooke's law states that stress is proportional to strain. Therefore,

$$\sigma_x = E\epsilon_x = -\frac{Ey}{\rho} \tag{d}$$

The significance of the subscript x on ϵ and σ is that the normal strain ϵ_x and the corresponding normal stress σ_x are in the x direction. The importance of Eq. (d) is that the normal stress is directly proportional to the distance y from the neutral axis of the section. If y is zero, the stress is zero; if y is large, the stress is large.

Equation (d) is also useful because we can use it to find the location of the neutral axis for any beam section. Since the beam is in pure bending, a summation of the forces in the x direction on any section must be zero. An element of area dA of the section (Fig. 5-6b) will have acting upon it a differential force of magnitude

$$dF_x = \sigma_x dA = -\frac{Ey}{\rho} \, dA \tag{e}$$

If we sum these forces across the entire section and state that they must add to zero, we have

$$\Sigma F_x = \int \sigma_x \, dA = -\frac{E}{\rho} \int y \, dA = 0 \tag{f}$$

The term

$$\int y \, dA = 0 \tag{g}$$

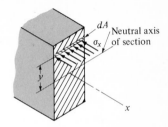

FIGURE 5-7.

of Eq. (f) states that "the summation of the moments of all the differential areas of the cross section about the neutral axis is zero." This means that the neutral axis of the section is the same as the centroidal axis of the section. (See Sec. 7 of Appendix A-7.) Thus, *the neutral axis of the beam is coincident with the centroid of the section.*

Equation (d) is a formula for the stress, and Eq. (g) locates the neutral axis so that we know the origin of the coordinate y. Our next problem is to get a relation between the bending moment and the stress, because the stress results from the application of the moment.

Recall in the previous chapter that a distinction was made between external and internal bending moments. These two are exactly equal and they balance each other to maintain static equilibrium. In Fig. 5-7, the normal stress σ_x acting on an element of area dA yields a force

$$dF = \sigma_x \, dA$$

The moment of this force about the neutral axis of the section is

$$dM_{\text{int}} = y \, dF = y\sigma_x \, dA \tag{h}$$

Equation (h) gives the differential internal moment. We get the total internal moment by integrating across the entire cross section. Since this is the same as the external moment, we can write

$$M_{\text{ext}} = \int dM_{\text{int}} = \int y\sigma_x \, dA \tag{i}$$

If we now substitute Eq. (f) into (i) and designate the external moment simply as M, we get

$$M = \frac{E}{\rho} \int y^2 \, dA \tag{j}$$

The integral in Eq. (j) is identified as the *area moment of inertia*, also called *second moment of area*, of the section about the centroidal axis. It is designated as I, as noted in Appendix A-8, and is

$$I = \int y^2 \, dA \tag{5-1}$$

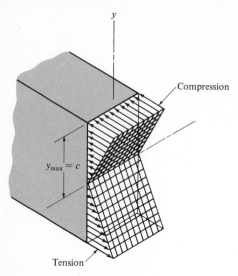

FIGURE 5-8. Distribution of bending stress on a section of a rectangular beam.

Thus, $M = EI/\rho$ or, as it is usually written,

$$\frac{1}{\rho} = \frac{M}{EI} \tag{5-2}$$

Since M, E, and I are all constants for a beam in pure bending, using the idealizations already mentioned, Eq. (5-2) states that the beam is bent into an arc of a circle, because the radius of curvature ρ is constant. This is also an important equation to be used in a future chapter for the analysis of beam deflections. By eliminating the radius of curvature from Eqs. (*d*) and (5-2), we finally get

$$\sigma_x = -\frac{My}{I} \tag{5-3}$$

which is called the *flexure formula*. It states that the normal stress, also called the *bending stress,* is directly proportional to the bending moment M and to the distance y from the neutral axis. The negative sign indicates that this stress is compression when y is positive and tension when y is negative, provided, of course, the bending moment is positive. Such a distribution is shown for a rectangular section in Fig. 5-8.

The maximum bending stress occurs when y is maximum. This will occur on the outer surface of the beam. Figure 5-8 shows that we designate this distance as c. If we omit the subscript x and the minus sign in Eq. (5-3), we can express the maximum bending stress as

$$\sigma = \frac{Mc}{I} \tag{5-4}$$

Equation (5-4) is often written in two alternative forms:

$$\sigma = \frac{M}{I/c} \qquad \sigma = \frac{M}{Z} \tag{5-5}$$

where $Z = I/c$ is called the *section modulus.*

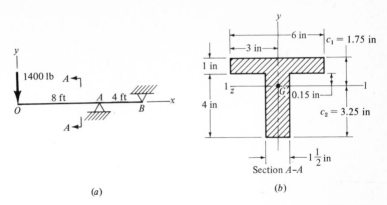

FIGURE 5-9.

EXAMPLE 5-1 Figure 5-9*a* shows a steel beam with a load of 1400 lb acting on the overhanging end. The beam is simply supported at *A* and *B*. The cross section of the beam, shown in Fig. 5-9*b*, is the same T section used in Example 5 of Appendix A-8. As shown, the centroid *G* was found to be 1.75 in below the top of the T. The moment of inertia was found to be 27.25 in⁴ about axis 1-1 through *G*. Note that this now becomes the neutral axis of the section and that the *x* axis of the beam is a line coming vertically out of the paper at *G*, normal to the *y* and *z* axes.

(a) Draw the loading, shear-force, and bending-moment diagrams and calculate the maximum bending moment. Neglect the weight of the beam.

(b) Find the radius of curvature and the location of the center of curvature of the beam corresponding to the maximum moment.

(c) Plot a stress-distribution diagram for the section corresponding to the maximum bending moment and compute the stresses on top and bottom and where the top of the T joins the stem.

SOLUTION

(a) The loading, shear-force, and bending-moment diagrams are derived using the methods of the previous chapter. They are shown in Fig. 5-10. Notice that the maximum moment is negative and that it occurs at support *A*.

(b) We can get the radius of curvature by using Eq. (5-2) and the maximum bending moment. Note that the bending moment must be expressed in inch units to be consistent with the units of *E* and *I* in Eq. (5-2). From Table A-9, we find $E = 3.0(10)^7$ psi for steel. Therefore, the radius of curvature of the beam at *A* is

$$\rho = \frac{EI}{M} = \frac{3.0(10)^7(27.25)}{-(11\ 200)(12)} = -6080 \text{ in} \quad Ans.$$

The minus sign means that the center of curvature is measured in the negative *y* direction. Thus, the coordinates of the center of curvature are

$$x = 8(12) = 96 \text{ in} \quad y = -6080 \text{ in} \quad Ans.$$

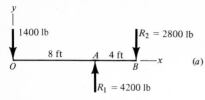

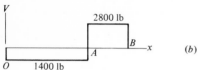

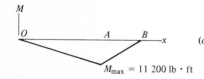

$M_{max} = 11\ 200\ lb \cdot ft$

(c)   **FIGURE 5-10.** (*a*) **Loading diagram;** (*b*) **shear-force diagram;** (*c*) **bending-moment diagram.**

(**c**) From Fig. 5-9*b*, we see that the distances from the neutral axis to the top and bottom surfaces of the beam are $c_1 = 1.75$ in and $c_2 = 3.25$ in. Using Eq. (5-4) and noting that the top surface is in tension, we find the stress to be

$$\sigma = \frac{Mc_1}{I} = \frac{(11\ 200)(12)(1.75)}{27.25} = 8\ 600\ psi \quad Ans.$$

The bottom surface is in compression. The stress is

$$\sigma = \frac{Mc_2}{I} = -\frac{(11\ 200)(12)(3.25)}{27.25} = -16\ 000\ psi \quad Ans.$$

The stem joins the T bar at $y = +0.75$ in, as shown in Fig. 5-9*b*. Using Eq. (5-3) and noting that the stress is tension, we get

$$\sigma = \frac{My}{I} = \frac{(11\ 200)(12)(0.75)}{27.25} = 3\ 690\ psi \quad Ans.$$

The resulting stress distribution across the section is shown in Fig. 5-11. Note that the stress at any other section along the length of the beam would be less because the bending moment, as shown in Fig. 5-10*c*, is less.   ////

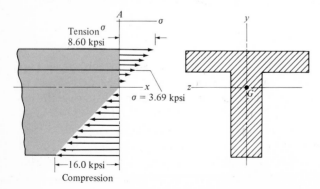

FIGURE 5-11.

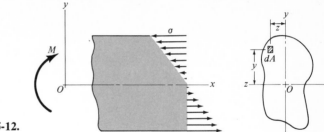

FIGURE 5-12.

5-3. BEAMS WITH UNSYMMETRICAL SECTIONS   The relations developed in Sec. 5-2 can also be applied to beams having unsymmetrical sections, provided that the plane of bending coincides with one of the two principal axes of the section. We have found that the stress at a distance y from the neutral axis is

$$\sigma = -\frac{Ey}{\rho} \tag{a}$$

Therefore, the force on the element of area dA in Fig. 5-12 is

$$dF = \sigma \, dA = -\frac{Ey}{\rho} \, dA$$

Taking moments of this force about the y axis and integrating across the section gives

$$M_y = \int z \, dF = \int \sigma z \, dA = -\frac{E}{\rho} \int yz \, dA \tag{b}$$

We recognize the last integral in Eq. (*b*) as the product of inertia I_{yz}. (See Appendix A-8, Sec. 7.) If the bending moment on the beam is in the plane of one of the principal axes, then

$$I_{yz} = \int yz \, dA = 0 \tag{c}$$

With this restriction, the relations developed in Sec. 5-2 hold for any cross-sectional shape. Of course, this means that the designer has a special responsibility to assure that the bending loads do, in fact, come onto the beam in a principal plane.

5-4. SHEAR STRESSES IN BEAMS   We learned in solving many of the problems in Chap. 4 that most beams have *both* shear forces and bending moments present. It is only occasionally that we encounter beams subjected to pure bending, that is to say, beams having zero shear force. And yet, the flexure formula was developed using the assumption of pure bending. As a matter of fact, the reason for assuming pure bending was simply to eliminate the complicating effects of shear force in the development. For engineering purposes, the flexure formula is valid no matter whether a shear force is present or not. For

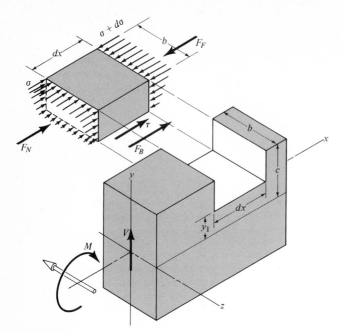

FIGURE 5-13.

this reason, we shall utilize the same normal bending-stress distribution [Eqs. (5-3) and (5-4)] when shear forces are present too.

In Fig. 5-13, we show a beam of constant cross section subjected to a shear force V and a bending moment M. The direction of the bending moment is easier to visualize by associating the hollow vector with your right hand. The hollow vector points in the negative z direction. If you will place the thumb of your right hand in the negative z direction, then your fingers, when bent, will indicate the direction of the moment M. By Eq. (4-11), the relationship of V to M is

$$V = \frac{dM}{dx} \qquad (a)$$

At some point along the beam, we cut a transverse section of length dx down to a distance y_1 above the neutral axis, as illustrated. We remove this section to study the forces which act. Because a shear force is present, the bending moment is changing as we move along the x axis. Thus, we can designate the bending moment as M on the near side of the section and as $M + dM$ on the far side. The moment M produces a normal stress σ and the moment $M + dM$ a normal stress $\sigma + d\sigma$, as shown. These normal stresses produce normal forces on the vertical faces of the element, the compressive force on the far side being greater than on the near side. The resultant of these two would cause the section to tend to slide in the $-x$ direction, and so this resultant must be balanced by a shear force acting in the $+x$ direction on the bottom of the section. This shear force results in a shear stress τ, as shown. Thus, there are three resultant forces acting on the element: F_N, due to σ, acts on the near face; F_F, due to $\sigma + d\sigma$, acts on the far face: and F_B, due to τ, acts on the bottom face. Let us evaluate these forces.

For the near face, select an element of area dA. The stress acting on this area is σ, and so the force is the stress times the area, or σdA. The force acting on the entire near face is the sum of all the σdA's, or

$$F_N = \int_{y_1}^{c} \sigma \, dA \qquad (b)$$

where the limits indicate that we integrate from the bottom $y = y_1$ to the top $y = c$. Using $\sigma = My/I$ [Eq. (5-3)], Eq. (b) becomes

$$F_N = \frac{M}{I} \int_{y_1}^{c} y \, dA \qquad (c)$$

The force on the far face is found in a similar manner. It is

$$F_F = \int_{y_1}^{c} (\sigma + d\sigma) \, dA = \frac{M + dM}{I} \int_{y_1}^{c} y \, dA \qquad (d)$$

The force on the bottom face is the shear stress τ times the area of the bottom face. Since this area is $b \, dx$, we have

$$F_B = \tau b \, dx \qquad (e)$$

Summing these three forces in the x direction gives

$$\Sigma F_x = +F_N - F_F + F_B = 0 \qquad (f)$$

If we substitute Eqs. (c) and (d) for F_N and F_F, and solve the result for F_B, we get

$$F_B = F_F - F_N = \frac{M + dM}{I} \int_{y_1}^{c} y \, dA - \frac{M}{I} \int_{y_1}^{c} y \, dA = \frac{dM}{I} \int_{y_1}^{c} y \, dA \qquad (g)$$

Next, using Eq. (e) for F_B, and solving for the shear stress gives

$$\tau = \frac{dM}{dx} \frac{1}{Ib} \int_{y_1}^{c} y \, dA \qquad (h)$$

Then, by the use of Eq. (a), we finally get the shear-stress formula as

$$\tau = \frac{V}{Ib} \int_{y_1}^{c} y \, dA \qquad (5\text{-}6)$$

In this equation, the integral is the first moment of the area of the vertical face about the neutral axis. This moment is usually designated as Q. Thus,

$$Q = \int_{y_1}^{c} y \, dA \qquad (5\text{-}7)$$

With this final simplification, Eq. (5-6) may be written as

$$\tau = \frac{VQ}{Ib} \qquad (5\text{-}8)$$

In using this equation, note that b is the width of the section at the particular distance y_1 from the neutral axis. Also, I is the moment of inertia of the entire section about the neutral axis.

5-5. SHEAR STRESS IN RECTANGULAR-SECTION BEAMS The purpose of this section is to show how the equations of the preceding section are used to find the shear-stress distribution in a beam having a rectangular cross section. Figure 5-14 shows a portion of a beam subjected to a shear force V and a bending moment M. As a result of the bending moment, a normal stress σ is developed on a cross section such as A-A, which is in compression above the neutral axis and in tension below. To investigate the shear stress at a distance y_1 above the neutral axis, we select an element of area dA at a distance y above the neutral axis. Then, $dA = b \, dy$, and so Eq. (5-7) becomes

$$Q = \int_{y_1}^{c} y \, dA = b \int_{y_1}^{c} y \, dy = \frac{by^2}{2} \Big|_{y_1}^{c} = \frac{b}{2} (c^2 - y_1^2) \qquad (a)$$

Substituting this value for Q into Eq. (5-8) gives

$$\tau = \frac{V}{2I} (c^2 - y_1^2) \qquad (5\text{-}9)$$

This is the general equation for the shear stress in a rectangular beam. To learn something about it, let us make some substitutions. From Table A-15, we learn that the moment of inertia for a rectangular section is $I = bh^3/12$; substituting $h = 2c$ and $A = bh = 2bc$ gives

$$I = \frac{Ac^2}{3} \qquad (b)$$

If we now use this value of I for Eq. (5-9), and rearrange, we get

$$\tau = \frac{3V}{2A} \left(1 - \frac{y_1^2}{c^2}\right) \qquad (5\text{-}10)$$

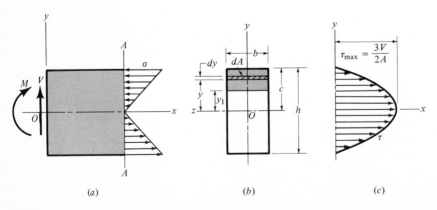

(a)   (b)   (c)

FIGURE 5-14.

TABLE 5-1

y_1	τ
0	1.500V/A
c/4	1.406V/A
c/2	1.125V/A
3c/4	0.656V/A
c	0

Now let us substitute various values of y_1, beginning with $y_1 = 0$ and ending with $y_1 = c$. The results are displayed in Table 5-1. We note that the maximum shear stress exists with $y_1 = 0$, which is at the neutral axis. Thus,

$$\tau_{max} = \frac{3V}{2A} \tag{5-11}$$

for a rectangular section. As we move away from the neutral axis, the shear stress decreases until it is zero at the outer surface where $y_1 = c$. This is a parabolic distribution and is shown in Fig. 5-14c. It is particularly interesting and significant here to observe that the shear stress is maximum at the neutral axis where the normal stress, due to bending, is zero, and that the shear stress is zero at the outer surfaces where the bending stress is a maximum.

5-6. THE CONCEPT OF A STRESS ELEMENT[1]   Figure 5-15a is the loading diagram of a simply supported beam loaded by a force F. Let us investigate in more detail the stress situation at some point such as A on the beam. To do this, we imagine an infinitely small prismatical element cut out of the stressed beam (Fig. 5-15b). This element is aligned with the xy reference system of the beam and has sides of length dx and dy and unit thickness. We have chosen element A between the reaction R_1 and the force F and on the neutral axis. The left side of the element will have a shear stress acting upward because R_1 acts upward. The right side will have a shear stress acting downward because F acts downward.

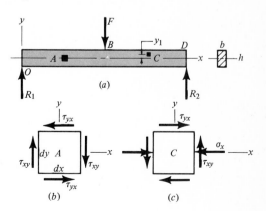

(a)

(b)

(c)

[1] See also Sec. 3-8 and Fig. 3-15.

FIGURE 5-15.

This pair of shear stresses, one on the left and one on the right, are equal and opposite and create a clockwise couple. Since the element is in static equilibrium, the couple must be balanced by another couple. This balancing couple comes from another pair of equal and opposite shear stresses acting on the top and bottom faces of the element shown in Fig. 5-15*b*.

Such an element is called a *stress element*. The faces of the element which are normal to the *x* axis are called the *x faces*. Those normal to the *y* axis are the *y faces*. The shear stresses acting on these faces have two subscripts; the first subscript indicates the face, the second subscript indicates the direction of the stress. Thus, τ_{xy} is a shear stress on an *x* face in the *y* direction, τ_{yx} is a shear stress on a *y* face in the *x* direction. These two pairs of stresses are, of course, equal to each other and, for a rectangular section, are given by Eq. (5-9) of the preceding section.

In this book, we shall not employ plus and minus signs for shear stresses. Instead, when using stress elements, we shall use clockwise (cw) and counterclockwise (ccw) to indicate directions. Thus, in Fig. 5-15*b*, the τ_{xy} stresses are cw and the τ_{yx} stresses are ccw.

A stress element located a distance y_1 from the neutral axis at *C* in Fig. 5-15*a*, will in addition have a normal stress $\sigma = My_1/I$, as shown in Fig. 5-15*c*. Note that this is designated as σ_x because it acts on the *x* face and is shown as a compressive stress. Note also that the shear stresses on the element at *C* have different directions than for the element at *A*. This results because the internal shear force *V* changes sign as we move from the left of a section at *B* to the right.

5-7. SHEAR FLOW   In determining the shear stress in a beam, the dimension *b* is not always measured parallel to the neutral axis. The beam sections shown in Fig. 5-16 show how to measure *b* in order to compute the static moment *Q*. Always remember that it is the tendency of the shaded area to slide relative to the unshaded area, which causes the shear stress. Thus, the shear stress at any section is always obtained by selecting the minimum *b* for that section.

Another useful idea in the analysis of beams is the concept of shear flow. By merely removing the dimension *b* from Eq. (5-8), we get the *shear flow q* as

$$q = \frac{VQ}{I} \tag{5-12}$$

FIGURE 5-16.

where q is in force units per unit of length of the beam at the section under consideration. Basically, the shear flow is simply shear force per unit length at the section defined by $y = y_1$. When the shear flow is known, the shear stress is determined by the equation

$$\tau = \frac{q}{b} \tag{5-13}$$

EXAMPLE 5-2 Figure 5-17 illustrates an aluminum box beam. It is built up by riveting two aluminum plates $\frac{1}{4}$ in thick to a pair of cold-formed 6 in by 1 in channels. The beam is subjected to a shear force of 1700 lb, which is constant along the length of the beam. If the $\frac{1}{4}$-in aluminum rivets used have an allowable shear load of 500 lb each, what should be the spacing a of the rivets. Neglect the bearing stresses.

SOLUTION

It is necessary first to find the moment of inertia of the built-up section. The easiest way to do this is to find the moment of inertia of the enclosing rectangle, and from this, to subtract the moments of inertia of the three empty rectangular spaces. The enclosing rectangle is 4 in wide and $6\frac{1}{2}$ in high. The central space is 2 in wide by 6 in high. The two side spaces are each $\frac{3}{4}$ in wide by $5\frac{1}{2}$ in high. Thus, the moment of inertia is

$$I = \frac{4(6\frac{1}{2})^3}{12} - \frac{2(6)^3}{12} - 2\left[\frac{(\frac{3}{4})(5\frac{1}{2})^3}{12}\right] = 34.7 \text{ in}^4$$

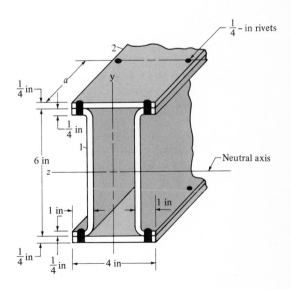

FIGURE 5-17.

We are going to solve this problem by using enough rivets to successfully resist the shear flow between the plate and the two channels. Thus, Q is the area of a plate times its centroidal distance. Therefore,

$$Q = A\bar{y} = (4)\left(\tfrac{1}{4}\right)\left(3\tfrac{1}{4} - \tfrac{1}{8}\right) = 3.125 \text{ in}^3$$

Then, using Eq. (5-12), we get

$$q = \frac{VQ}{I} = \frac{(1700)(3.125)}{34.7} = 153 \text{ lb/in}$$

This means that 153 lb of shear force is to be transferred from the plate to the channels for every inch of beam length. Each pair of rivets will transfer 1000 lb into the channels. Therefore, the rivet spacing must be

$$a = \frac{1000}{153} = 6.54 \text{ in} \quad Ans.$$

In other words, there must be a pair of rivets every $6\tfrac{1}{2}$ in along the length of the beam on both the top and the bottom. ////

EXAMPLE 5-3. The T-section beam shown in Fig. 5-18 is constructed by brazing two $\tfrac{1}{4}$ in by 2 in aluminum plates to an upright $\tfrac{1}{4}$ in by 5 in plate. If the beam is to carry a shear force of 800 lb, constant along its length, what is the shear stress in the brazing metal?

SOLUTION

We must first locate the neutral axis of the section. Taking moments about the lower end of the vertical plate, we have

$$\bar{y} = \frac{A_1y_1 + 2A_2y_2}{A_1 + 2A_2} = \frac{(5)\left(\tfrac{1}{4}\right)(2.5) + 2(2)\left(\tfrac{1}{4}\right)\left(5 - \tfrac{1}{8}\right)}{(5)\left(\tfrac{1}{4}\right) + 2(2)\left(\tfrac{1}{4}\right)} = 3.56 \text{ in}$$

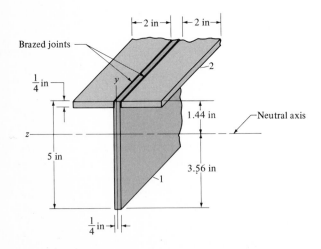

FIGURE 5-18.

which is shown in Fig. 5-18. Next, we find the moment of inertia to be

$$I = (I_1 + d_1^2 A_1) + 2(I_2 + d_2^2 A_2)$$

$$= \left[\frac{(\frac{1}{4})(5)^3}{12} + (3.56 - \frac{5}{2})^2 (5)(\frac{1}{4}) \right] + 2 \left[\frac{(2)(\frac{1}{4})^3}{12} + (1.44 - \frac{1}{8})^2 (2)(\frac{1}{4}) \right]$$

$$= 5.74 \text{ in}^4$$

Here we want to find the shear stress in the brazed joint. Thus, we are interested in learning how much shear force is transferred from the $\frac{1}{4}$ in by 2 in plate to the stem of the T. Thus,

$$Q = A_2 \bar{y}_2 = (\frac{1}{4})(2)(1.44 - \frac{1}{8}) = 0.657 \text{ in}^3$$

Then, using Eq. (5-12), we find the shear flow to be

$$q = \frac{VQ}{I} = \frac{800(0.657)}{5.74} = 92 \text{ lb/in}$$

which is the shear force transferred through the braze metal in every inch of beam length. Thus, the shear stress in the braze metal is

$$\tau = \frac{q}{b} = \frac{92}{1/4} = 368 \text{ psi} \qquad Ans. \qquad ////$$

5-8. OTHER SHEAR-STRESS DISTRIBUTIONS

The methods of Sec. 5-7 can be used to find the shear stress at any point in a beam section. So, in Fig. 5-19, we present the approximate distributions in a solid-round section, a tube, and an I section. For a solid-round section, the maximum shear stress is

$$\tau_{max} = \frac{4V}{3A} \tag{5-14}$$

For a tube

$$\tau_{max} = \frac{2V}{A} \tag{5-15}$$

If the web is assumed to take the entire shear load, then for a WF or I section beam, the approximate maximum shear stress is

$$\tau_{max} = \frac{V}{A_w} \tag{5-16}$$

where A_w is the area of the web. Of course, all of these occur at the neutral axis.

It should be noted that these shear stresses are completely neglected for most beams because they are so inconsequential. It is safer, however, to calculate these stresses until it is learned which ones can be ignored. Shear stresses may need to be considered for wood beams because wood has such a low shear strength. Also, watch out for beams that have very thin webs, some built-up beams, which we have already covered, and short beams. A short beam is one with a small length-to-depth ratio.

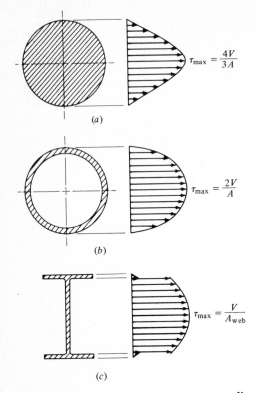

$$\tau_{max} = \frac{4V}{3A}$$

(a)

$$\tau_{max} = \frac{2V}{A}$$

(b)

$$\tau_{max} = \frac{V}{A_{web}}$$

(c)

FIGURE 5-19. Approximate shear-stress distributions for three beam sections.

5-9. SECONDARY SHEAR IN BOLTED AND RIVETED BEAMS Two or more bolts or rivets, or welding, are generally used whenever the reaction at a beam support is to include a moment. If the beam is fastened using rivets or bolts, then it may be that the forces exerted on these fasteners are not equal for all the fasteners in the group. If this should be the case, a different approach than that used in Chap. 3 would have to be developed.

Consider the cantilever of Fig. 5-20*a*. The beam is fastened to the column by four bolts—*A*, *B*, *C*, and *D*. It is loaded by a concentrated force *F* at the end at a distance *l* from the centroid of the bolt group. Usually, the centroid can be found by inspection. But if several bolt sizes are used, or if the spacing is unusual, then the methods of Appendix A-7, Sec. 2, should be employed. Ideally, the centroid of the bolt group should be on the principal axis of the column.

Figure 5-20*b* is a free-body diagram of the beam showing that the four bolts must transfer a shear force *V* and a bending moment *M* into the beam.

Figure 5-20*c* is an enlarged free-body diagram of the bolt group. Two force vectors are shown acting on each bolt. The shear force *V* is assumed to be divided equally among the four bolts. Thus,

$$F'_A = F'_B = F'_C = F'_D = \frac{V}{4} \qquad (a)$$

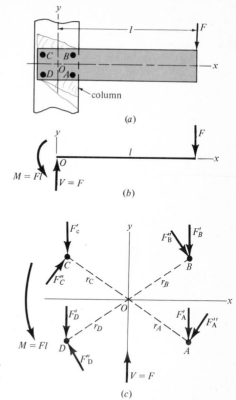

FIGURE 5-20. (*a*) Cantilever bolted to column; (*b*) free-body diagram of beam; (*c*) free-body diagram of bolts.

The force F' is called the *direct load* or the *primary shear*. For a connection with n bolts, the primary shear is given by the equation

$$F' = \frac{V}{n} \tag{5-17}$$

The resistance of the bolt to the moment load F'' is called the *secondary shear*. As shown, this produces an additional force on the bolt. To find its magnitude, we first designate the distances from the centroid of the bolt group to the bolt centers as r_A, r_B, etc., as shown in Fig. 5-20c. Taking a summation of moments about O gives

$$\Sigma M_O = M - F_A'' r_A - F_B'' r_B - F_C'' r_C - F_D'' r_D = 0 \tag{b}$$

In the general case, this equation will contain more than one unknown, and so another relation is necessary.

It is clear that if a bolt were located at the centroid O, it could not resist a moment. A bolt close to A will take a small moment; a bolt far from A will take a large moment. Thus, the secondary shear force F'' depends upon the distance r. Therefore,

$$\frac{F_A''}{r_A} = \frac{F_B''}{r_B} = \frac{F_C''}{r_C} = \frac{F_D''}{r_D} \tag{c}$$

If we now solve Eqs. (b) and (c) simultaneously for F_A'', say, we get

$$F_A'' = \frac{Mr_A}{r_A^2 + r_B^2 + r_C^2 + r_D^2} \tag{d}$$

or, more generally,

$$F_i'' = \frac{Mr_i}{\sum_0^n r^2} \tag{5-18}$$

This equation states that the secondary shear load on the ith bolt is equal to the bending moment times the distance from the centroid O to the center of the ith bolt, divided by the sum of the squares of the distances to all n bolts in the group.

The last step is to determine the resultants of the primary and secondary shear-force vectors. This can be done graphically or by trigonometry if the angles can be found. Many times, only the bolt or bolts having the largest resultant would be needed, since these would rule the analytical results. In Fig. 5-20c, bolts A and B will have the largest resultant. Once this resultant is found, the connection may be analyzed using the methods of Chap. 3, with one exception. When a beam is bolted or riveted, the strength of the member must be based, usually, on the bending stress. In many cases, the member is weakest across the bolt holes. The following example illustrates the significant differences.

EXAMPLE 5-4   Shown in Fig. 5-21 is a $\frac{5}{8}$ by $7\frac{1}{2}$-in rectangular steel bar cantilevered to a 10-in channel. The cantilever is fastened to the channel by four $\frac{5}{8}$-in bolts. Based on an external load of 3000 lb, find:
(a) The resultant load on each bolt
(b) The maximum bolt shear stress
(c) The maximum bearing stress
(d) The critical bending stress in the bar

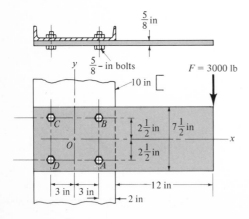

FIGURE 5-21.

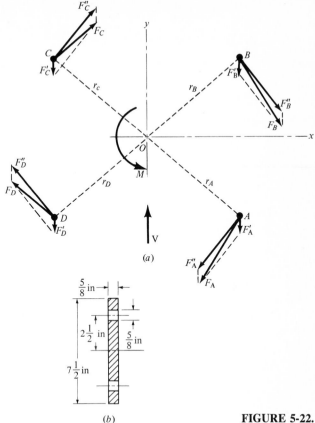

FIGURE 5-22.

SOLUTION

(a) Figure 5-22a is a free-body diagram of the bolt group, drawn to scale. The reactions are

$$V = 3000 \text{ lb} \qquad M = (3000)(12 + 2 + 3) = 51\ 000 \text{ lb} \cdot \text{in}$$

By inspection, it is seen that the centroid O of the bolt group is equidistant from the bolts. This distance is

$$r = \sqrt{(3)^2 + (2.5)^2} = 3.90 \text{ in}$$

The primary shear load per bolt is

$$F' = \frac{V}{n} = \frac{3000}{4} = 750 \text{ lb}$$

Since the secondary shear loads are equal, Eq. (5-18) becomes

$$F'' = \frac{Mr}{4r^2} = \frac{M}{4r} = \frac{51\ 000}{4(3.90)} = 3270 \text{ lb}$$

The primary and secondary shear forces are plotted to scale for each bolt in Fig. 5-22a, and the resultants obtained using the parallelogram rule. The magnitudes of the resultants for A and B are equal and greater than for C and D, although the directions are different for all four bolts. By measurement we find these to be

$F_A = F_B = 3890$ lb   *Ans.*

$F_C = F_D = 2750$ lb   *Ans.*

An alternate analytical method of obtaining these results is obtained by reference to Fig. 5-23. Here, for example, the angle that r_A makes with the x axis, is

$$\theta = \tan^{-1} \frac{2.5}{3} = 39.8°$$

Now, at bolt A, resolve the secondary shear F''_A into its x and y components. Since $F'_A = 750$ lb and $F''_A = 3270$ lb, the x and y components of the resultant force on bolt A are seen to be

$F_{Ax} = -3270 \sin 39.8° = -2100$ lb

$F_{Ay} = -750 - 3270 \cos 39.8° = -3270$ lb

The resultant bolt force is, therefore,

$F_A = \sqrt{(-2100)^2 + (-3270)^2} = 3890$ lb   *Ans.*

In a similar manner, using bolt D, we have

$F_{Dx} = -3270 \sin 39.8° = -2100$ lb

$F_{Dy} = -750 + 3270 \cos 39.8° = 1770$ lb

$F_D = \sqrt{(-2100)^2 + (1770)^2} = 2750$ lb   *Ans.*

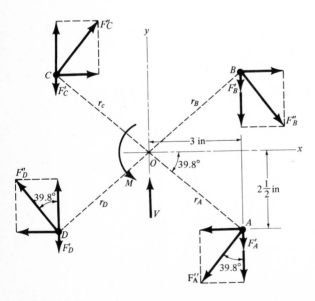

FIGURE 5-23.

(b) Bolts A and B are critical because they carry the largest shear load. The bolt shear-stress area is $A_s = \pi d^2/4 = \pi(0.625)^2/4 = 0.307$ in². The shear stress is

$$\tau = \frac{F}{A_s} = \frac{3890}{0.307} = 12\ 700\ \text{psi} \qquad Ans.$$

(c) The bearing-stress area is $A_b = td = (0.625)(0.625) = 0.390$ in². So the bearing stress is

$$\sigma = -\frac{F}{A_b} = -\frac{3890}{0.390} = -9970\ \text{psi} \qquad Ans.$$

(d) The critical bending stress in the bar is assumed to occur in a section parallel to the y axis and through bolts A and B. At this section, the bending moment is

$$M = 3000(12 + 2) = 42\ 000\ \text{lb} \cdot \text{in}$$

The moment of inertia through this section (Fig. 5-22b) is obtained by use of the transfer formula. Thus,

$$I = I_{\text{plate}} - 2(I_{\text{hole}} + d^2 A)$$

$$= \frac{0.625(7.5)^3}{12} - 2\left[\frac{0.625(0.625)^3}{12} + (2.5)^2(0.625)(0.625)\right]$$

$$= 17.1\ \text{in}^4$$

Then,

$$\sigma = \frac{Mc}{I} = \frac{(42\ 000)(7.5/2)}{17.1} = 9200\ \text{psi} \qquad Ans. \qquad ////$$

PROBLEMS

SECTION 5-2

5-1 to 5-6 For each section illustrated, find the moment of inertia, the location of the neutral axis, and the distances from the neutral axis to the top and bottom surfaces. Find also the stresses resulting from the application of a positive bending moment of 10 000 lb · in at the top and bottom surfaces, and at every abrupt change in cross section.

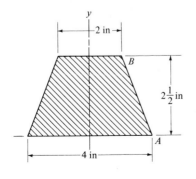

(a)

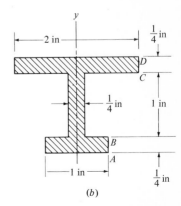

(b)

PROB. 5-1.

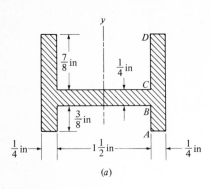

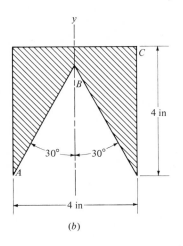

PROB. 5-2.

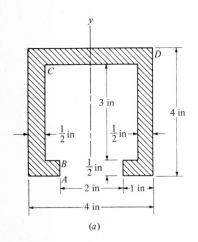

PROB. 5-3.

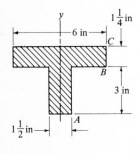

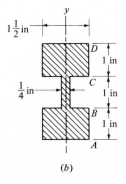

PROB. 5-4.

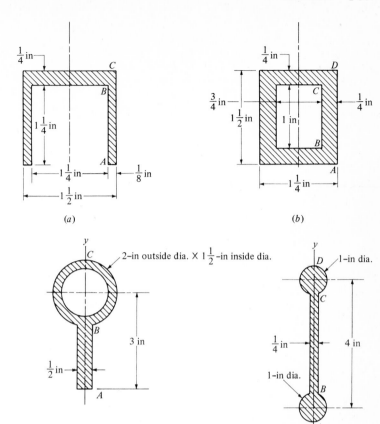

PROB. 5-5.   (a)   (b)

PROB. 5-6.   (a)   (b)

5-7 to 5-9   The figures illustrate a collection of composite shapes obtained
by riveting standard rolled shapes together. The properties of individual shapes can be obtained from the appropriate tables in Appendix A-10 to A-14. In each case, find the location of the neutral
axis, the distance or distances from the neutral axis to the outer surfaces, and the moment of inertia of the composite. In Prob. 5-8*a*,
neglect the effect of the filler plate between the two channels.

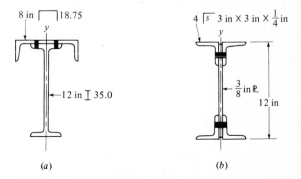

PROB. 5-7.   (a)   (b)

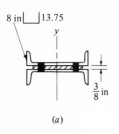

8 in ⌐ 13.75

y

$\frac{3}{8}$ in

(a)

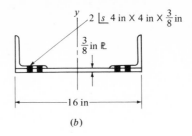

2 ⌐s⌐ 4 in × 4 in × $\frac{3}{8}$ in

y

$\frac{3}{8}$ in ℞

16 in

(b)

PROB. 5-8.

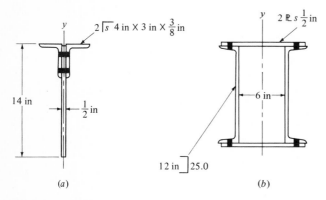

y   2 ⌐s⌐ 4 in × 3 in × $\frac{3}{8}$ in

14 in $\frac{1}{2}$ in

(a)

y   2 ℞ s $\frac{1}{2}$ in

6 in

12 in ⌐ 25.0

(b)

PROB. 5-9.

5-10 and 5-11   Determine the *xy* coordinates of the center of curvature corresponding to the place where the beam is bent the most, for each beam shown in the figure. The beams are all made of douglas fir (see Appendix A-9), and all have rectangular sections, as shown.

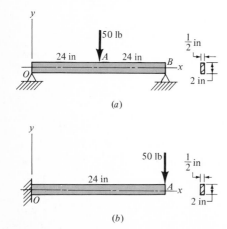

y

50 lb

24 in   *A*   24 in

B —*x*

$\frac{1}{2}$ in

2 in

O

(a)

y

50 lb

24 in

$\frac{1}{2}$ in

A —*x*

2 in

O

(b)

PROB. 5-10.

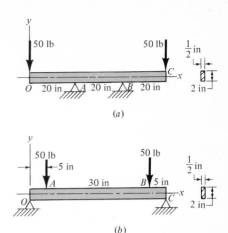

PROB. 5-11.

5-12   The allowable tensile stress in beams made of structural steel is 20 kpsi. In each case below, the length of the beam is between simple supports. Considering the weight of the beams, find the maximum safe uniformly distributed external load that each beam can carry.

(a) A 14 WF 84 beam 16 ft long

(b) A beam made of 6- by 6- by $\frac{1}{2}$-in angles, back-to-back, 10 ft long. The plane of bending is along the back-to-back direction

(c) A beam made of two 10-in 30-lb channels, back-to-back (backs in bending plane) 22 ft long

(d) A 20 I 75 beam, 18 ft long

5-13   The same as Prob. 5-12 for the following beams:

(a) A 6-in tube with $\frac{3}{8}$-in wall thickness, 12 ft long

(b) A 36 WF 260 beam, 40 ft long

(c) A 27 I 120 beam, 24 ft long

(d) A beam made of two 6 by 4 by $\frac{1}{2}$ in angles with the long legs riveted back-to-back and in the bending plane, 6-ft long

SECTION 5-5

5-14 and 5-15   For each beam illustrated, find the locations and magnitudes of the maximum tensile bending stress and the maximum shear stress.

SECTION 5-7

5-16   A wood beam is built by nailing four 1 × 2's to a 2 × 12 as shown in the figure. Since lumber is sized prior to dressing, the actual dimensions are shown in parentheses. The 1 × 2's are nailed to the 2 × 12 using 10d nails. (10d means ten penny and refers to the length of the nail. A 10d common nail is, very closely, 0.128 in in diameter and 3 in long.) If the shear force in the beam is 1200 lb and a 10d nail will take a shear load of 140 lb, what should be the spacing of the nails?

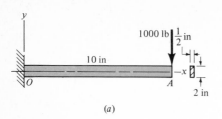

(a)

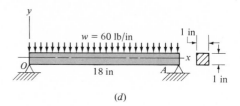

(b)

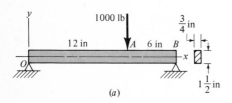

(c)

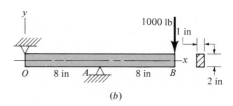

(d)

PROB. 5-14.

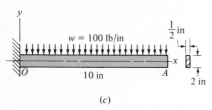

(a)

(b)

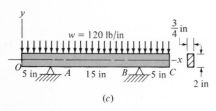

(c)

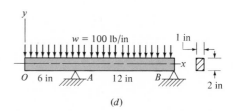

(d)

PROB. 5-15.

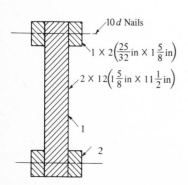

PROB. 5-16.

5-17 The same as Prob. 5-16, except that the section shown for this problem uses $16d$ nails and the allowable shear load per nail is 160 lb.

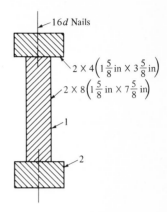

$16d$ Nails

$2 \times 4 \left(1\frac{5}{8}\text{ in} \times 3\frac{5}{8}\text{ in}\right)$

$2 \times 8 \left(1\frac{5}{8}\text{ in} \times 7\frac{5}{8}\text{ in}\right)$

1

2

PROB. 5-17.

5-18 A T-section beam is built up by riveting two angles to a plate, as shown. If the allowable shear load for the $\frac{1}{2}$ in rivets is 3000 lb per rivet for single shear, what should be the rivet spacing? The beam shear load is $V = 6200$ lb.

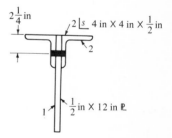

$2\frac{1}{4}$ in

$2\; \underline{\llcorner s}\; 4$ in $\times$ 4 in $\times \frac{1}{2}$ in

2

$\frac{1}{2}$ in $\times$ 12 in ℝ

1

PROB. 5-18.

5-19 A T-section beam is built up by riveting a channel to an I beam, as shown in the figure. The shear force in the beam is 9100 lb. If $\frac{1}{2}$ in rivets are used, having an allowable shear load of 3000 lb per rivet, what spacing should be used?

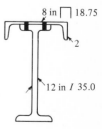

8 in ⌐⌐ 18.75

2

12 in I 35.0

PROB. 5-19.

5-20 The figure shows a beam made by spot welding two strips on to back-to-back cold-formed channels. Determine the spacing of each pair of welds if the shear force on the beam is 900 lb and if each spot weld can take a shear load of 1500 lb.

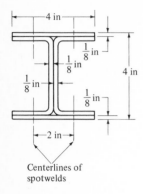

Centerlines of
spotwelds

PROB. 5-20.

5-21 The figure shows a box beam made by gluing four boards, each finished to 1 by 6 in, together. The beam is to support a shear force of 1650 lb.
   (a) Find the shear stress in the joints if bending takes place in the xy plane.
   (b) Find the shear stress in the joints if bending takes place in the xz plane.

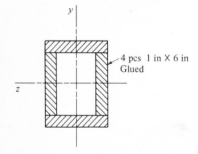

4 pcs 1 in × 6 in
Glued

PROB. 5-21.

SECTION 5-9
   5-22 to 5-24 For each of the connections shown, find the maximum bolt shear stress and the maximum bending stress in the beam.

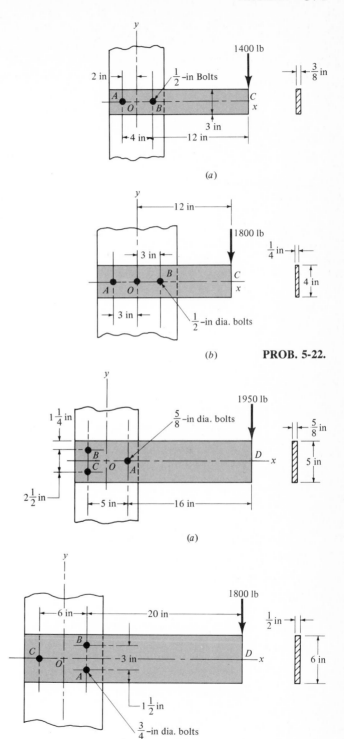

(a)

(b)   **PROB. 5-22.**

(a)

(b)   **PROB. 5-23.**

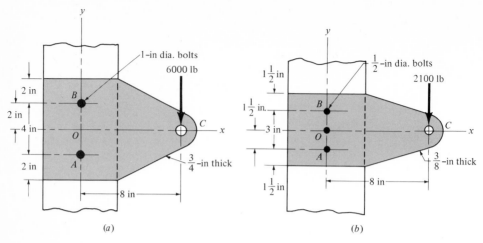

(a)   (b)

PROB. 5-24.

SI UNITS

5-25 For each section shown in the figure, find the moment of inertia, the location of the neutral axis, and the distances from the neutral axis to the outer surfaces. Assume that these are sections of a beam subjected to a positive bending moment of $1000 \text{ N} \cdot \text{m}$. Find the stresses at the outer surfaces.

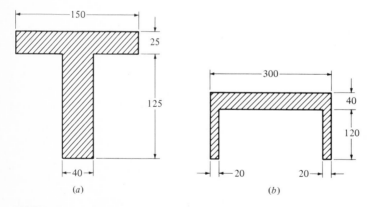

(a)   (b)

PROB. 5-25.   **All dimensions in millimetres.**

5-26 Determine the maximum load F in kilonewtons that each of the beams shown in the figure can take if the bolt shear stress is not to exceed 100 MPa, and the bending stress if the member is not to exceed 80 MPa.

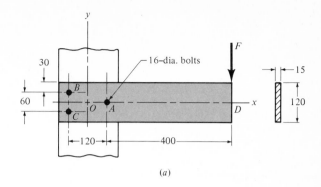

(a)

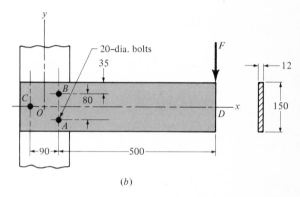

(b)

PROB. 5-26. **All dimensions in millimetres.**

CHAPTER 6

torsion

6-1. TORSIONAL DEFORMATION   A moment that causes twisting of a bar or a rod is called *torsion*. In Fig. 6-1, the external twisting moment T, also called the external *torque,* may be designated by drawing arrows on the surface to indicate the direction, or by the hollow vectors on the x axis in the sense of the right-hand rule.

As we shall see, it is necessary to restrict our consideration of torsion to round bars. These, however, may be either solid or hollow.

Due to the torque, a line AC originally parallel to x will be twisted through the angle γ to AC'. The corresponding radius OC rotates through θ to OC'. The basic assumption in the theory of torsion is that the straight line OC remains straight during the deformation.

We can also visualize the deformation by examining an element of the shaft. The element, originally located at B, moves to B' during twisting and is deformed from a square to a parallelogram by the shear stresses that act on the sides.

The shear deformation is the distance CC', which is equal to $r\theta$. But the *shear strain* is the angle γ. Since $\gamma l = r\theta$, we see that

$$\gamma = \frac{r\theta}{l} \tag{a}$$

which shows that the shear strain varies directly with the radius r.

In Fig. 6-2, we select an element from inside the shaft of length dx and radius ρ. We replace θ by $d\theta$ and write Eq. (a) as

$$\gamma = \rho \frac{d\theta}{dx} \tag{b}$$

If we now consider the shaft cross section in the yz plane, say, then the resultant of the stress distribution must equal the applied moment T because of the rotational symmetry of this distribution. A stress element of area dA will have a force exerted on it of $\tau \, dA$, and the moment of this force is $(\tau \, dA)\rho$. The sum of these moments is the same as the external torque, and so

$$T = \int (\tau \, dA)\rho \tag{c}$$

If we write Hooke's law for shear (that stress is proportional to strain), we have $\tau = G\gamma$, where G is the modulus of elasticity in shear. With Eq. (b), this relation becomes

$$\tau = G\rho \frac{d\theta}{dx} \tag{d}$$

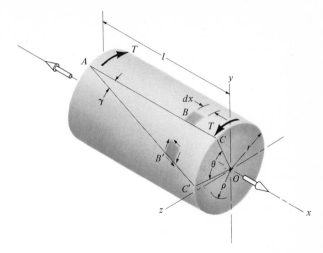

FIGURE 6-1.

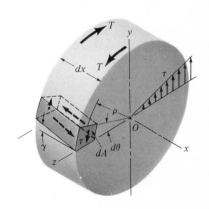

FIGURE 6-2.

Substituting Eq. (*d*) into (*c*) gives

$$T = G \frac{d\theta}{dx} \int \rho^2 \, dA \qquad (e)$$

From Appendix A-8, we recognize the integral as the polar moment of inertia. Designating this as

$$J = \int \rho^2 \, dA \qquad (f)$$

enables us to solve Eq. (*e*) for the twist per unit length. Thus,

$$\frac{d\theta}{dx} = \frac{T}{GJ} \qquad (g)$$

If we multiply both sides of Eq. (*g*) by *dx* and integrate between the limits of $x = 0$ and $x = l$, we get the angle of twist of a shaft having a constant cross section and a length *l*. The result is

$$\theta = \frac{Tl}{GJ} \qquad (6\text{-}1)$$

where θ is the total angle of twist in radians.

To obtain the shear stress, eliminate the term $d\theta/dx$ from Eqs. (*d*) and (*g*) and solve for τ. The result is

$$\tau = \frac{T\rho}{J} \qquad (6\text{-}2)$$

This is the distribution, proportional to the radius ρ, which is shown in Fig. 6-2. Of course, the stress will reach a maximum on the outer surface of the shaft with $\rho = r$. Thus,

$$\tau_{max} = \frac{Tr}{J} \qquad (6\text{-}3)$$

6-2. TORSIONAL STRESS   Using Table A-15, we find the polar moment of inertia for a solid round bar to be

$$J = \frac{\pi d^4}{32} \qquad (6\text{-}4)$$

By substituting this formula for *J* into Eq. (6-3), we get the special formula for the maximum shear stress in a solid round shaft as

$$\tau_{max} = \frac{16T}{\pi d^3} \qquad (6\text{-}5)$$

Using Table A-15 again, we find for tubing

$$J = \frac{\pi}{32} (D^4 - d^4) \qquad (6\text{-}6)$$

where *D* and *d* are the outside and inside diameters, respectively.

In computing shear stresses in rotating shafts, it is often necessary to obtain the torque *T* from a consideration of the horsepower transmitted and the shaft velocity. Based upon the fact that one horsepower is equivalent to 33 000 foot-pounds of energy per minute, three relations can be stated as follows:

$$P = \frac{2\pi Tn}{33\ 000(12)} \qquad (6\text{-}7)$$

$$P = \frac{FV}{33\ 000} \qquad (6\text{-}8)$$

$$T = \frac{63\ 000P}{n} \qquad (6\text{-}9)$$

where P = horsepower transmitted along the shaft
   T = torque transmitted, lb · in
   n = shaft speed, rpm
   F = tangential force at radius r, lb
   V = tangential velocity at radius r, ft/min

EXAMPLE 6-1   Figure 6-3 shows a geared steel countershaft. Gear A, having 80 teeth, is driven by a 20-tooth gear mounted on an 80-horsepower (hp) motor that runs at 1120 rpm. This 80 hp is transmitted along the shaft to gear B where 50 hp is used in driving another machine, not shown. The remaining power is transmitted to gear C and then to a third machine that is also not shown. Note the symbol used to designate a shaft bearing. By supporting the shaft on these bearings, very little bending takes place, and so we can neglect the bending stresses. Find the maximum shear stress in the shaft and the torsional deflections between A and B, B and C, and A and C.

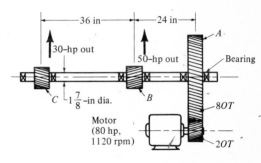

FIGURE 6-3.

SOLUTION

Because of the tooth numbers, gear A will make one turn for every four turns of the motor. Therefore, the countershaft speed is

$$n = \frac{20}{80}(1120) = 280 \text{ rpm}$$

Using Eq. (6-9), we find the torque exerted by gear A on the countershaft will be

$$T = \frac{63\ 000\ P}{n} = \frac{(63\ 000)\ (80)}{280} = 18\ 000 \text{ lb · in}$$

This is the torsion in the shaft between A and B. Using Eq. (6-5), we find the maximum shearing stress to be

$$\tau_{max} = \frac{16T}{\pi d^3} = \frac{(16)\ (18\ 000)}{\pi (1.875)^3} = 13\ 900 \text{ psi}   \quad Ans.$$

Next, using Table A-9, we find the modulus of rigidity for steel to be $G = 11.5$ Mpsi. Using Eq. (6-4) for polar moment of inertia and Eq. (6-1), we find the formula for the torsional deflection of a solid round shaft is

$$\theta = \frac{Tl}{GJ} = \frac{Tl}{G(\pi d^4/32)} = \frac{32Tl}{\pi d^4 G} \tag{6-10}$$

Then the deflection of the shaft between gear A and gear B is

$$\theta = \frac{32(18\ 000)(24)}{\pi(1.875)^4(11.5)(10)^6} = 0.031 \text{ rad} \qquad Ans.$$

Since there are π rad in $180°$, the angle in degrees is

$$\theta = 0.031 \frac{180}{\pi} = 1.78° \qquad Ans.$$

Since 50 hp is used at gear B, only three-eighths (30/80) of the total torque is transmitted from B to C. This amounts to

$$T = \frac{3}{8}(18\ 000) = 6750 \text{ lb} \cdot \text{in}$$

Thus, the torsional deflection between B and C is, from Eq. (6-10),

$$\theta = \frac{32Tl}{\pi d^4 G} = \frac{32(6750)(36)}{\pi(1.875)^4(11.5)(10)^6} = 0.017 \text{ rad} \qquad Ans.$$

Or, in degrees,

$$\theta = 0.017 \frac{180}{\pi} = 1.00° \qquad Ans.$$

Then the deflection of the shaft from A to C is

$$\theta = 1.78 + 1.00 = 2.78° \qquad Ans. \qquad\qquad ////$$

6-3. TORSIONAL STRENGTH   A brief discussion of methods of testing materials in direct shear is included in Chap. 2 (Sec. 2-6). Shear strengths can also be determined using torsion tests of materials. On a torsion-test machine, methods are available for measuring the torque T and the angle of twist θ over a definite gauge length l. Measurements are carried out from the beginning of twist until the bar ruptures. The results are then plotted as a *torque-twist diagram.* Such diagrams have θ as the abscissa and T as the ordinate, and hence they resemble ordinary stress-strain diagrams. On these diagrams, the curve will begin to deviate from a straight line at the proportional limit. Since this is roughly the same as the yield point, we can designate this torque as T_y. By substituting this torque into Eq. (6-3), we have

$$S_{sy} = \frac{T_y r}{J} \tag{6-11}$$

where S_{sy} is called the *torsional yield strength,* and it roughly corresponds to the *yield strength in shear.*

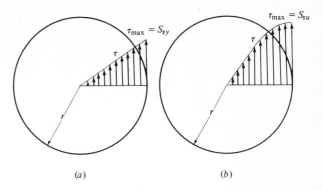

(a)   (b)

FIGURE 6-4. Stress distribution during a torque-twist test.

Figure 6-4*a* shows the stress distribution in the torsion-test specimen at the instant $T = T_y$. Note, especially, that it is only the outer surface of the specimen that has reached the yield point. Thus, the outer surface has reached the limit of elasticity, and the core or remainder of the bar is still in the elastic state.

Now, if we continue to increase the torque T, eventually the outer surface will reach the *ultimate torsional strength* S_{su} and begin to rupture. The stress distribution will then resemble that shown in Fig. 6-4*b*. If we designate the corresponding torque as T_u and substitute it into Eq. (6-3) as before, we have

$$S_{su} = \frac{T_u r}{J} \tag{6-12}$$

But note, very particularly, that Eq. (6-3) was developed under the assumption that Hooke's law applies, i.e., stress is proportional to strain. In Fig. 6-4*b*, the outer third or so of the cross-sectional area has a curved stress distribution. Therefore, stress is *not* proportional to strain in this area. It is for this reason that Eq. (6-12) is not exactly correct. To emphasize this fact, the strength S_{su}, obtained from this equation, is properly called the *modulus of rupture*. So S_{su} should *not* be called the ultimate shear strength.

6-4. THE PRINCIPAL STRESSES IN TORSION   Figure 6-5*a* shows a solid round bar twisted by a torque T. Because of this torsion, a stress element at A, aligned with the three reference axes, will have shear stresses acting on it as given by Eqs. (6-2) and (6-3). In Fig. 6-5*b*, we imagine that we remove this element from the bar for closer examination. Here we see that the side normal to x has the shear stress τ_{xy}. We have already learned that the first subscript indicates the face and the second the direction. Thus, τ_{yx} on the y face acts in the x direction. Also, since this stress element is in static equilibrium, $\tau_{xy} = \tau_{yx}$.

In order that we can deal with force equilibrium in this section, we have also made the stress element a cube, with the sides of the cube one unit in

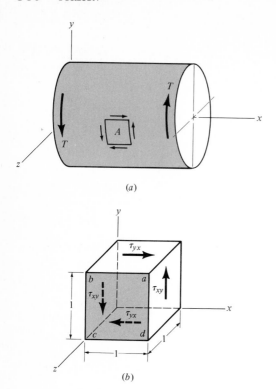

(a)

(b)

FIGURE 6-5.

length, as illustrated in Fig. 6-5*b*. This makes it possible to compute the *shear forces* on the faces of the element. Thus, the shear force on the *x* face is

$$F_y = \tau_{xy}A = \tau_{xy}(1)(1) = \tau_{xy} \quad \text{pounds}$$

where A is the area of the face. Note that the shear force on the left *x* face is negative, for example.

Next, we wish to cut this stress element into two parts on 45° diagonals, as shown in Fig. 6-6*a* and *b*. Note that the depths are not shown because we have already specified that they are to have unit depth. Now, in order to maintain force equilibrium on the two halves of the element, unknown stresses σ and τ must act on the diagonal faces. The directions of these are assumed as shown. Using the pythagorean theorem, we find the lengths of the diagonals *ac* and *bd* to be $\sqrt{2}$ units. Next, we resolve the stresses σ and τ into rectangular components parallel to the *x* and *y* axes, as shown. We can use the pythagorean theorem again to find these components because they are equal to each other. Thus, for example, the stress σ resolves into a stress component $\sigma/\sqrt{2}$ parallel to *x* and another stress component $\sigma/\sqrt{2}$ parallel to *y*.

Now refer specifically to Fig. 6-6*a* and sum the *forces*, not stresses, in the *x* direction and equate them to zero. This means we must multiply the stresses by the areas. To make the computation clear, we show the areas in brackets. Thus,

$$\Sigma F_x = -\tau_{yx}[(1)(1)] + \frac{\tau}{\sqrt{2}}[\sqrt{2}(1)] + \frac{\sigma}{\sqrt{2}}[\sqrt{2}(1)] = 0 \qquad (a)$$

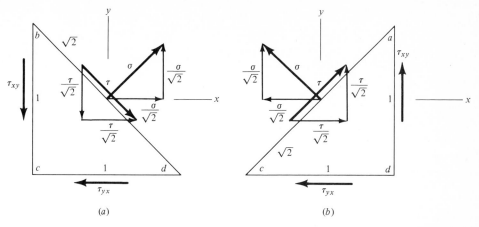

(a)   (b)

FIGURE 6-6.

And, for the y direction,

$$\Sigma F_y = -\tau_{xy}[(1)(1)] - \frac{\tau}{\sqrt{2}}[\sqrt{2}(1)] + \frac{\sigma}{\sqrt{2}}[\sqrt{2}(1)] = 0 \qquad (b)$$

If we next simplify both expressions and solve them for σ, we get the pair of simultaneous equations

$$\sigma = \tau_{yx} - \tau$$
$$\sigma = \tau_{xy} + \tau \qquad (c)$$

Remembering that $\tau_{xy} = \tau_{yx}$, we can add this pair of equations together, eliminating τ, and find that

$$\sigma = \tau_{xy} \qquad (d)$$

Substituting this value of σ in either of Eqs. (c), we learn that $\tau = 0$.

The normal stress given by Eq. (d) is called a *principal stress* because it occurs on a face having zero shear stress. Thus, whenever a stress element is oriented such that the shear stresses on the faces of the element are zero, then the normal stresses acting on these same faces are called the *principal stresses*. For an element in pure torsion, two principal stresses exist — one on diagonal *bd* and another on diagonal *ac*. Because of the importance of this development, we rewrite Eq. (d) as

$$\sigma_1 = \tau_{xy} \qquad (6\text{-}13)$$

in which the subscript 1 designates the first principal, most positive, stress. By subjecting Fig. 6-6b to the same analysis, we learn that the other principal stress is

$$\sigma_2 = -\tau_{xy} \qquad (6\text{-}14)$$

and is compressive.

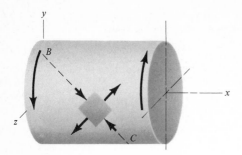

FIGURE 6-7.

If we now draw a new stress element at the same location as A in Fig. 6-5a, but oriented 45° from the xy system, then we get the stress state shown in Fig. 6-7. This is a characteristic of pure torsion—tension in a 45° angle in one direction, and compression in the other. Pay particular attention to the direction of the torque T. If this torque is reversed, then the principal stresses are reversed.

Brittle materials are known to fail in tension rather than in compression. A brittle material torqued as in Fig. 6-7 would then fail along the helix BC. Try it by twisting a piece of blackboard chalk, and note the helix angle when it breaks.

6-5. HELICAL SPRINGS The design and analysis of rotating shafts provides, as we have learned, an excellent example of the use of torsion theory. Another good example is provided by the tension- or compression-loaded helical spring. One can visualize the torsion in a spring by picturing in the mind's eye a coiled garden hose. Now pull one end of the hose along the axis of the coil. As each turn is pulled off, the hose twists about its own axis. And so the flexing of a helical spring in tension or compression creates torsion in the wire in a similar manner.

Figure 6-8a shows a round-wire helical compression spring. Compression springs usually have the ends ground flat, or perhaps squared during winding, to transfer the load into the spring more uniformly. Tension springs have hook ends to transfer the load into the body of the spring. In this example, we designate the *wire diameter* as d and the *mean coil diameter* as D. If the *outside spring diameter D_o* is given, then the mean diameter is

$$D = D_o - d \qquad\qquad (a)$$

To determine the stress in the wire, imagine that a portion of the spring is removed (Fig. 6-8b) and its effect replaced by the internal forces, thus creating a free-body diagram of the remaining portion. As shown, a force F acts at the cut portion of the wire. The twin forces F, equal, opposite, and parallel, form a cw couple whose arm is $D/2$. To balance this couple requires that a ccw torque $T = FD/2$ be applied at the cut section too. The stresses in the wire are due,

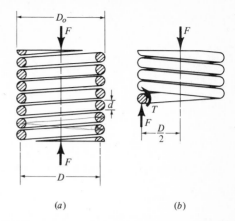

(a)   (b)

FIGURE 6-8. (*a*) **A helical spring loaded in compression,** (*b*) **free-body diagram of portion of spring showing that the wire is subjected to a direct shear and a torsional shear.**

therefore, to these two separate effects, the torsion T and the direct shear force F. Using Eq. (6-2), the torsional stress component is

$$\tau = \frac{T\rho}{J} = \frac{FD\rho}{2J} \qquad (b)$$

where the stress τ is computed for any radius ρ from zero to $d/2$. The distribution of this stress is shown in Fig. 6-9*a*.

The shear stress component due to the force F is

$$\tau = \frac{F}{A} \qquad (c)$$

and is uniform across the diameter d, as shown in Fig. 6-9*b*.

As shown in Fig. 6-9*c*, the maximum shear stress occurs on the inside of the spring where the two stress distributions are additive. Using $\rho = d/2$, $J = \pi d^4/32$, and $A = \pi d^2/4$ for Eqs. (*b*) and (*c*), and adding them, gives

$$\tau_{max} = \frac{8FD}{\pi d^3} + \frac{4F}{\pi d^2} \qquad (6\text{-}15)$$

for the stress on the inside of the coil. The same equation applies for tension springs.

To obtain the equation for the deflection, we consider an element of wire of length dx cut from the spring (Fig. 6-10). A line ab on the surface of the wire will rotate to ac through the angle γ when the spring is deformed. Neglecting the direct shear stress and writing Hooke's law,

$$\tau = \frac{8FD}{\pi d^3} = G\gamma$$

or $\quad \gamma = \dfrac{8FD}{\pi d^3 G} \qquad (d)$

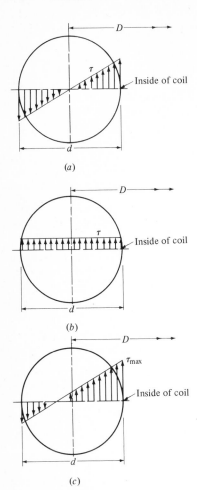

(a)

(b)

(c)

FIGURE 6-9. (*a*) **Stress distribution due to torsion alone;** (*b*) **stress distribution due to direct shear alone;** (*c*) **resultant stress distribution due to both torsion and direct shear.**

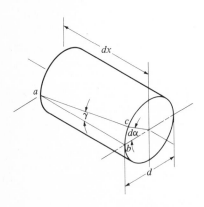

FIGURE 6-10.

The distance bc in Fig. 6-10 is $\gamma\, dx$, and the differential angle $d\alpha$, through which one end rotates with respect to the other, is

$$d\alpha = \frac{\gamma\, dx}{d/2} = \frac{2\gamma\, dx}{d} \qquad (e)$$

If we designate l as the total length of the wire in the stressed portion of the spring, then

$$\alpha = \frac{2\gamma}{d} \int_0^l dx = \frac{2\gamma l}{d} \qquad (f)$$

which is the angular deflection or twist of the wire. The linear deflection of the spring is equal to the angular deflection of the wire times the mean radius of the spring; this follows from the definition of radian measure. Denoting this linear deflection as y gives

$$y = \alpha \frac{D}{2} = \frac{\gamma l D}{d} \qquad (g)$$

In order to get Eq. (g) into a more usable form, let $l = \pi D N$, where N is the number of *active* or useful *coils*. At the ends of the spring, where the wire contacts the support, no spring action is possible. For this reason, these coils are said to be *dead* or *inactive*. Most compression coil springs will have one-half to one dead coil at each end, depending upon how the ends are formed in manufacture. If we now substitute l in terms of N and Eq. (d) for γ into Eq. (g), we finally get

$$y = \frac{\gamma l D}{d} = \left(\frac{8FD}{\pi d^3 G}\right)(\pi D N)\left(\frac{D}{d}\right) = \frac{8FD^3N}{d^4 G} \qquad (6\text{-}16)$$

for the spring deflection.

Another useful concept is spring rate, or spring constant. *Spring rate* is the spring force per unit of deflection, or

$$k = \frac{F}{y} \qquad (6\text{-}17)$$

Substituting y from Eq. (6-16) gives another formula in terms of the geometry and spring material as

$$k = \frac{d^4 G}{8D^3 N} \qquad (6\text{-}18)$$

EXAMPLE 6-2   A 22-gauge (0.028 in) extension spring, made of hard-drawn steel wire, has an outside diameter of $\frac{1}{2}$ in. The spring has 48 active coils, hook ends, and is wound with an initial tension of 0.15 lb. With no external load, the distance between hooks is $2\frac{1}{4}$ in.
(a) Determine the spring rate.

(b) The torsional yield strength of 22-gauge hard-drawn spring wire is 118 kpsi. What external force would be required to initiate yielding and hence cause a permanent set in the spring?

(c) What is the extended length of the spring when yielding begins?

SOLUTION

(a) The mean spring diameter is

$$D = D_o - d = 0.5 - 0.028 = 0.472 \text{ in}$$

Using $G = 11.5$ Mpsi for steel and Eq. (6-18), the spring rate is

$$k = \frac{d^4 G}{8 D^3 N} = \frac{(0.028)^4 (11.5)(10)^6}{(8)(0.472)^3(48)} = 0.175 \text{ lb/in} \qquad Ans.$$

(b) We first rearrange Eq. (6-15) as follows:

$$\tau_{max} = \frac{8FD}{\pi d^3} + \frac{4F}{\pi d^2} = \frac{8FD}{\pi d^3}\left(1 + \frac{d}{2D}\right) \qquad (6\text{-}19)$$

Next, substitute S_{sy} for τ_{max} and solve the result for the force F. We get

$$F = \frac{\pi d^3 S_{sy}}{8D}\frac{1}{1 + (d/2D)} = \frac{\pi (0.028)^3(118)(10)^3}{(8)(0.472)}\frac{1}{1 + [0.028/(2)(0.472)]}$$
$$= 2.093 \text{ lb} \qquad Ans.$$

(c) The initial tension or preload is given as 0.15 lb and the spring is closed solid with a length between hooks of $2\frac{1}{4}$ in. Thus, 0.15 lb must be applied before the spring begins to extend. Therefore, an additional force of

$$F = 2.093 - 0.15 = 1.943 \text{ lb}$$

causes the spring to extend until yielding begins. From Eq. (6-17), the deflection required is

$$y = \frac{F}{k} = \frac{1.943}{0.175} = 11.1 \text{ in}$$

Therefore, when yielding begins, the distance between hook ends will be

$$L = 2.25 + 11.1 = 13.35 \text{ in} \qquad Ans. \qquad \text{////}$$

6-6. TORSION OF RECTANGULAR SECTIONS The analysis of the stresses and deflections caused by the twisting of nonround shapes is a problem involving the solution of partial differential equations of elasticity; it is too advanced for inclusion in this book. Consequently, we present here only the results for the analysis of rectangular sections.[1]

In Fig. 6-11, a rectangular bar of thickness t, width w, and length l is twisted by a moment T acting about the x axis in accordance with the right-hand

[1] See W. Flugge, "Handbook of Engineering Mechanics," p. 36-7, McGraw-Hill Book Company, New York, 1962.

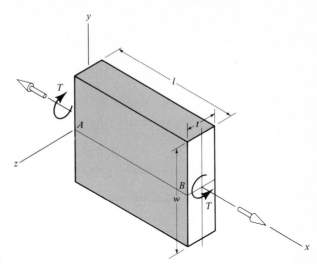

FIGURE 6-11.

TABLE 6-1
Torsion constants for rectangular sections.

w/t	1.0	1.2	1.5	2.0	2.5	3	4	5	10	∞
c_1	4.81	4.57	4.33	4.06	3.88	3.74	3.55	3.44	3.20	3.00
c_2	7.11	6.02	5.10	4.37	4.02	3.80	3.56	3.44	3.20	3.00

rule. The shear stresses at the corners of the bar are zero. The maximum shear stress occurs along line AB at the middle of the longest of the two cross-sectional dimensions. This stress is

$$\tau_{\max} = c_1 \frac{T}{wt^2} \tag{6-20}$$

The torsional deflection of the bar is

$$\theta = c_2 \frac{Tl}{Gwt^3} \tag{6-21}$$

The constants c_1 and c_2 are given in Table 6-1.

6-7. TORSION IN WELDED JOINTS   Figure 6-12 illustrates a cantilever of length l welded to a column by three fillet welds. We have learned that the reaction at the support of a cantilever consists of a shear force and a moment. The shear force produces a primary shear in the welds of magnitude

$$\tau' = \frac{F}{A} \tag{a}$$

where A is the throat area of all three welds.

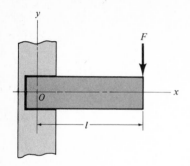

FIGURE 6-12.

The moment at the support produces secondary shear or torsion of the welds, and this stress is

$$\tau'' = \frac{Tc}{J} \qquad (b)$$

where c is the distance from the centroid of the weld group to the point in the weld of interest, and J is the polar moment of inertia of the weld group about the centroid of the group. When the size of the welds are known, these equations can be solved and the results combined to obtain the maximum weld shear stress.

The reverse procedure is that in which the allowable shear stress is given and we wish to find the weld size. The usual procedure would be to estimate the weld size, compute J and A, and then find and combine τ' and τ''. If the resulting maximum shear stress were too large, a larger weld size would be estimated and the procedure would be repeated. Eventually, a satisfactory result would be obtained.

A much more useful approach to the problem is to treat each fillet weld as a line. The resulting polar moment of inertia is then equivalent to a *unit polar moment of inertia*. The advantage of treating the weld as a line is that the unit polar moment of inertia is the same regardless of the weld size. If we specify the leg of a fillet weld as h, then the throat width is $0.707h$ (see Sec. 3-7). Designating the unit polar moment of inertia as J_u, we have

$$J = 0.707hJ_u \qquad (6\text{-}22)$$

Here, J_u is found by the use of Eq. (4) in Appendix A-8, in which the width of the line is taken as unity. The transfer formula is also used, then, to find the unit polar moment of inertia of weld groups. Table 6-2 lists the throat area and unit polar moment of inertia for most of the fillet-weld configurations that we shall encounter in this book. The example that follows is illustrative of the calculations normally made.

EXAMPLE 6-3   A 12 000 lb load is transferred from a welded fitting into an 8 in, 11.5 lb channel (as illustrated in Fig. 6-13). Compute the maximum stress in the weld.

TABLE 6-2
Torsional properties of fillet welds. G is the centroid of the weld group; h is the weld size.

weld	throat area	location of G	unit polar moment of inertia
	$A = 0.707hd$	$\bar{x} = 0$ $\bar{y} = \dfrac{d}{2}$	$J_u = \dfrac{d^3}{12}$
	$A = 1.414hd$	$\bar{x} = \dfrac{b}{2}$ $\bar{y} = \dfrac{d}{2}$	$J_u = \dfrac{d(3b^2 + d^2)}{6}$
	$A = 0.707h(b + d)$	$\bar{x} = \dfrac{b^2}{2(b + d)}$ $\bar{y} = \dfrac{d^2}{2(b + d)}$	$J_u = \dfrac{(b + d)^4 - 6b^2d^2}{12(b + d)}$
	$A = 0.707h(2b + d)$	$\bar{x} = \dfrac{b^2}{2b + d}$ $\bar{y} = \dfrac{d}{2}$	$J_u = \dfrac{8b^3 + 6bd^2 + d^3}{12} - \dfrac{b^4}{2b + d}$
	$A = 1.414h(b + d)$	$\bar{x} = \dfrac{b}{2}$ $\bar{y} = \dfrac{d}{2}$	$J_u = \dfrac{(b + d)^3}{6}$
	$A = 1.414\pi hr$		$J_u = 2\pi r^3$

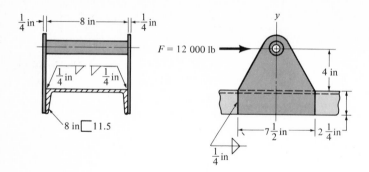

FIGURE 6-13.

SOLUTION

As shown by the figure, each plate is welded to the channel using three $\frac{1}{4}$ in fillet welds. We shall divide the load in half and consider only a single plate in the analysis to follow. Two of the three welds are $2\frac{1}{4}$ in long, and the length of the third is $7\frac{1}{2}$ in. Using Table 6-2, we first locate the centroid, also called center of gravity, of the weld group.

$$\bar{y} = \frac{b^2}{2b+d} = \frac{(2.25)^2}{2(2.25)+7.5} = 0.422 \text{ in}$$

$$\bar{x} = \frac{d}{2} = \frac{7.5}{2} = 3.75 \text{ in}$$

These dimensions are shown on the free-body diagram of Fig. 6-14a. Note that this also locates the origin O and the x axis. The reaction moment (or torque) is

$$M = T = (6000)(4.422) = 26\,500 \text{ lb} \cdot \text{in}$$

per plate.

Next, we refer to Table 6-2 and find the unit polar moment of inertia to be

$$J_u = \frac{8b^3 + 6bd^2 + d^3}{12} - \frac{b^4}{2b+d}$$

$$= \frac{(8)(2.25)^3 + (6)(2.25)(7.5)^2 + (7.5)^3}{12} - \frac{(2.25)^4}{(2)(2.25)+7.5}$$

$$= 104 \text{ in}^3$$

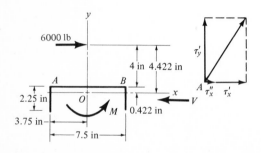

(a)

(b)   **FIGURE 6-14.**

Then, from Eq. (6-22),

$$J = 0.707hJ_u = (0.707)(0.25)(104) = 18.4 \text{ in}^4$$

Using Table 6-2 again, we find the throat area for the welds on one plate to be

$$A = 0.707h(2b + d) = (0.707)(0.25)[(2)(2.25) + 7.5] = 2.12 \text{ in}^2$$

The primary shear stress is

$$\tau'_x = \frac{F}{A} = \frac{6000}{2.12} = 2830 \text{ psi}$$

We find the secondary shear stress in components. The y component is

$$\tau''_y = \frac{Tc_x}{J} = \frac{(26\ 500)(3.75)}{18.4} = 5400 \text{ psi}$$

The x component is

$$\tau''_x = \frac{Tc_y}{J} = \frac{(26\ 500)(0.422)}{18.4} = 610 \text{ psi}$$

These stress components may be combined to give the maximum stresses, which occur at corners A and B. Thus, for A (Fig. 6-14b) we have

$$\tau = \sqrt{\tau_y^2 + \tau_x^2} = \sqrt{(5400)^2 + (2830 + 610)^2} = 6400 \text{ psi} \qquad Ans. \qquad ////$$

PROBLEMS

SECTION 6-2

6-1 Using a maximum allowable shear stress of 8000 psi, find the shaft diameter needed to transmit 50 hp when
(a) The shaft speed is 2000 rpm.
(b) The shaft speed is 200 rpm.

6-2 A $\frac{1}{2}$-in diameter steel bar is to be used as a torsion spring. If the torsional stress in the bar is not to exceed 16 kpsi when one end is twisted through an angle of 30°, what must be the length of the bar?

6-3 A 3-in diameter solid steel shaft, used as a torque transmitter, is replaced with a 3-in diameter hollow shaft having a $\frac{1}{4}$-in wall thickness. If both materials have the same strength, what is the percent reduction in torque transmission? What is the percent reduction in shaft weight?

6-4 A hollow steel shaft is to transmit 4000 lb · ft of torque and is to be sized such that the torsional stress does not exceed 22 kpsi.
(a) If the inside diameter is three-fourths of the outside diameter, what size shaft should be used?
(b) What is the stress on the inside of the shaft when full torque is applied?

6-5 A round steel bar is $\frac{3}{8}$ in in diameter and 12 ft long. One end of the bar is clamped and the other is twisted through one-half turn.
   (a) What is the maximum torsional stress in the bar?
   (b) If this bar were to be twisted by using a wrench 24 in in length, what force must be applied at the end of the wrench?

6-6 A 10-gauge steel music wire has a decimal diameter of 0.024 in. If such a wire is 6 ft long, what torque is required to twist it through three complete turns? What maximum torsional stress would be developed?

6-7 A hollow steel shaft is $3\frac{1}{2}$ in in diameter and has a $\frac{1}{4}$ in wall thickness. Find the shear stress on the outside and on the inside when the torque is 6000 lb · ft.

6-8 A $1\frac{1}{2}$-in diameter solid steel shaft transmits 65 hp at a speed of 2200 rpm. Find the maximum shear stress in the shaft. What would the shear stress be if the speed were doubled, but the same horsepower transmitted?

6-9 A $1\frac{1}{4}$-in diameter steel torsion bar has a $\frac{3}{8}$-in diameter hole bored out of the center for experimental purposes. In order to determine whether the bar is seriously weakened, compute the percent reduction in torsional strength due to drilling the hole.

6-10 A bar is in pure torsion if it is being twisted and nothing else. That is, there must be no bending nor axial loading. The figure shows a pipe that is to be tightened (torqued) into a fixture whose purpose is not shown. What force or forces should be applied to the wrench to assure that the pipe is in pure torsion? Show on a sketch how the forces are applied to the wrench.

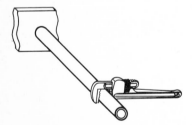

PROB. 6-10.

6-11 The geared steel countershaft *ABCD*, which is shown in the figure, is adequately supported by bearings so that bending stresses are neg-

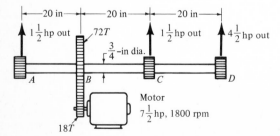

PROB. 6-11.

ligible. Based on the horsepower transmitted to the shaft by the motor, find the maximum torsional stress and the torsional deflection from A to B and from B to D.

SECTION 6-4

6-12 Draw a stress element having sides one unit in length and oriented to an xy system of axes. Let $\tau_{xy} = 10$ kpsi cw. Draw inclined faces whose normals are, respectively, 30, 45, and 75° ccw from the x axis, and find the shear stress and the normal stress on these faces. Make sketches to show the directions.

6-13 The same as Prob. 6-12, except the normals are 120, 135, and 165° ccw from the x axis.

SECTION 6-5

6-14 A helical compression spring is wound from 0.120-in diameter music wire. The spring has 12 total coils (11 of them active) and an outside diameter of 1 in.
  (a) Compute the spring rate.
  (b) Determine the free length of the spring such that when it is squeezed until the coils touch each other, the stress will not be greater than 110 kpsi.

6-15 A helical compression spring is wound using 0.072-in diameter music wire. The spring has a mean diameter of $\frac{3}{4}$ in, eight active coils, and one dead coil. The free length of the spring is 2.5 in and the torsional yield strength is 128 kpsi.
  (a) Find the spring rate.
  (b) Compression springs should be designed such that when they are squeezed solid, called the solid height or solid length, the stress will not exceed the torsional yield strength. Otherwise, if a person accidentally closed the spring, a permanent deformation would result and the free length would be incorrect for the original purpose. So compute the stress in the spring when it is closed solid, and compare the result with the strength.

6-16 A helical compression spring is made of 0.047 in wire having a torsional yield strength of 108 kpsi. It has an outside diameter of $\frac{1}{2}$ in and 14 active coils.
  (a) What force F will cause the material to yield?
  (b) What deflection would be caused by the load in (a)?
  (c) Calculate the scale of the spring.
  (d) If the spring has one dead turn at each end, what is the solid height?
  (e) What should be the free length of the spring such that when it is compressed solid, the material will not develop a permanent set?

6-17 A helical tension spring is made of 0.047 in wire having a torsional yield strength of 108 kpsi. The spring has an outside diameter of $\frac{1}{2}$ in, 36 active coils, and hook ends. The spring is prestressed to 10 kpsi during winding, which keeps it closed solid until an external tensile load is applied. (This is considered good practice in the manufacture of tension springs.) When closed solid, the distance between hook ends is $2\frac{3}{4}$ in.

(a) What is the preload in the spring?

(b) What tensile load would cause the material to yield?

(c) What is the spring rate?

(d) What would be the extended length of the spring at the beginning of yield?

SECTION 6-7

6-18 The tubing shown in the figure is welded to its support by welding all around using a $\frac{3}{8}$-in fillet weld, as indicated. If the allowable shear stress in the weld is 12 kpsi, what torque T can the joint withstand?

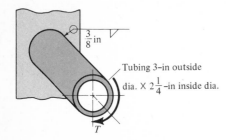

$\frac{3}{8}$ in

Tubing 3-in outside dia. $\times$ $2\frac{1}{4}$-in inside dia.

T

PROB. 6-18.

6-19 An angle is welded to one side of a column using a $\frac{1}{4}$-in fillet weld, according to the figure. Find the maximum stress in the weld and its location. The 500 lb bending force is assumed to act along a principal axis of the angle.

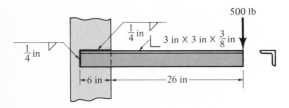

500 lb

$\frac{1}{4}$ in

3 in $\times$ 3 in $\times$ $\frac{3}{8}$ in

$\frac{1}{4}$ in

6 in   26 in

PROB. 6-19.

6-20 A plate supporting a load of 12 000 lb is welded to a column with $\frac{1}{2}$-in fillet welds, as indicated. Find the maximum stress in the welds and indicate where it acts.

6-21 The figure shows a typical beam connection as fabricated in a structural steel shop. The 21-in wide flange beam has four angles welded to its web, two at each end. This welding is done in the shop, after which the beam is shipped to the construction site for erection. In this problem, each angle is welded using three $\frac{3}{8}$-in fillet welds according to the figure. Also, the end connection is to withstand a shear reaction of $V = 23\ 000$ lb and a moment reaction of $M = 16\ 000$ lb · ft for each angle. Determine the maximum stress in the welds and specify where this occurs.

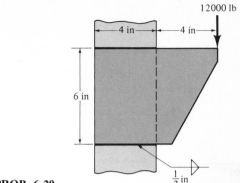

PROB. 6-20.

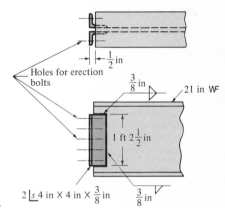

PROB. 6-21.

6-22 The figure illustrates a cantilevered bar loaded by a bending force of 3000 lb. The bar is to be welded top and bottom to a column 6 in in width. What size welds should be employed if the allowable stress in the welds is 16 000 psi?

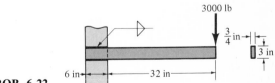

PROB. 6-22.

6-23 The column bracket shown in the figure consists of two 3 in angles to support the load, a $\frac{3}{4}$ in plate, and two 4 by 3 in angles welded to the plate by fillet welds along the toes of the 3 in sides. The finished bracket will be bolted to the column. The allowable stress in the welds is 8.0 kpsi. What size welds should be used?

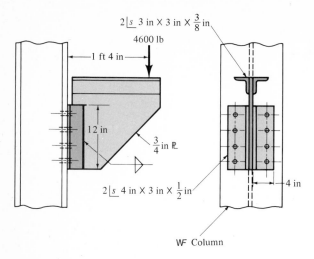

$2\lfloor s$ 3 in × 3 in × $\frac{3}{8}$ in

4600 lb

1 ft 4 in

12 in

$\frac{3}{4}$ in R.

$2\lfloor s$ 4 in × 3 in × $\frac{1}{2}$ in

4 in

W⁼ Column

PROB. 6-23.

SI UNITS

6-24 Using a maximum allowable shear stress of 60 MPa, find the shaft diameter needed to transmit a power of 40 kW when

(a) The shaft speed is 30 s⁻¹.*

(b) The shaft speed is 3 s⁻¹.

6-25 A round steel bar is 25 mm in diameter and 3.6 m long. It is twisted by a torque of 400 N · m. Find the angular deflection of the shaft and the maximum shear stress.

---

* In SI, the abbreviation for rotational frequency in revolutions per second is s⁻¹. This does not conflict with angular velocity, which is measured in radian per second and is abbreviated rad/s.

deflection of beams

7-1. INTRODUCTION The basic problem in this chapter is to determine the shape of the beam after it has been bent. As an example, Fig. 7-1 shows an overhanging beam, originally straight, bent by the concentrated force F. The deflection of the beam is given by the coordinate y, which is measured from the x axis. Note that the x axis does not bend with the beam, but remains in the same straight line it occupied before the bending force was applied. The coordinate y is a function of x and gives the position of the deflected center line of the beam from its original position. Thus, at $x = x_1$, a positive deflection $y = y_1$ occurs; at $x = x_2$, the deflection y_2 is negative.

It is worth noting that the coordinate y used in this chapter is different from the one used in Chap. 5. In Figs. 5-6 and 5-7, for example, y is the distance from the neutral axis of the section to an element of area in the section. In those figures, the maximum value of y is the distance from the neutral axis to the outer surface of the beam. But in this chapter, we are using y to give us the position of the neutral axis from the x axis after bending has taken place.

7-2. SUCCESSIVE DIFFERENTIATION AND CURVATURE In Secs. 4-3 and 4-4, we learned that the instantaneous rate of change, or slope, of a variable y with respect to another variable x is called the derivative of y with respect to x. Thus, when $y = f(x)$, then the slope of y with respect to x is found by evaluating the derivative dy/dx. If we designate this slope as y', then

$$y' = \frac{dy}{dx} \tag{a}$$

In exactly the same manner, it is possible to find the slope of y' with respect to x. If we designate this as y'', then

$$y'' = \frac{dy'}{dx} \tag{b}$$

Now, because of Eq. (a), (b) can be written

$$y'' = \frac{d}{dx}\left(\frac{dy}{dx}\right) \tag{c}$$

Equation (c) is called a *successive derivative* or the *second derivative* of y with respect to x. The physical significance has not changed. The new variable y'' is

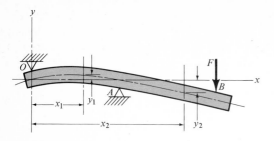

FIGURE 7-1.

simply the slope of the slope of y with respect to x; this is the same as the rate of change of the slope. The more usual notation is

$$y'' = \frac{d^2y}{dx^2} \tag{d}$$

which is read: y'' equals the second derivative of y with respect to x. Equation (*d*) *does not* mean that d in the numerator is squared, nor that x in the denominator is squared. The process indicated by Eqs. (*c*) and (*d*) can be continued for as long as desired. Thus,

$$y''' = \frac{d^3y}{dx^3} \qquad y^{iv} = \frac{d^4y}{dx^4} \qquad \text{etc.}$$

One of the more useful applications of successive differentiation is to be found in the study of curvature in the calculus. The *curvature K* of a plane curve $y = f(x)$ is defined as the reciprocal of the radius of curvature ρ. Thus,

$$K = \frac{1}{\rho} \tag{7-1}$$

In Example 5-1, we found the radius of curvature, at one point, of a beam 12 ft long to be 6080 in. So the curvature at this point is

$$K = \frac{1}{6080} = 164(10)^{-6} \text{ in}^{-1}$$

The exact formula for curvature, which we shall not derive, is

$$K = \frac{\dfrac{d^2y}{dx^2}}{\left[1 + \left(\dfrac{dy}{dx}\right)^2\right]^{3/2}} \tag{7-2}$$

This formula is difficult to use in the analysis of the curvature of beams because of the radical in the denominator. Note, from Fig. 7-1, that the term dy/dx in the denominator of Eq. (7-2) is actually the slope of the deflection curve. The slope is the angle that tangents to various points on the deflection curve make with the x axis, and it is measured in radians. But beam slopes are very very small, almost always less than 5°. A 5° angle is 0.087 radians. And $(0.087)^2 = 7.6(10)^{-3}$, which, compared to unity, is very small. Thus, except for beams with extremely large deflections, the denominator of Eq. (7-2) can be taken as unity. By combining Eq. (7-1) with (7-2), we then get for the radius of curvature of a beam

$$\frac{1}{\rho} = \frac{d^2y}{dx^2} \tag{7-3}$$

7-3. THE COMPLETE BEAM EQUATIONS   In Chap. 4 [Eq. (4-10)], the relation between shear force and loading was found to be

$$q = \frac{dV}{dx} \tag{7-4}$$

And we also learned [Eq. (4-11)] that the shear force was equal to the derivative of the bending moment with respect to x. Thus,

$$V = \frac{dM}{dx} \tag{7-5}$$

Our remaining task is to relate loading, shear force, and bending moment to the slope and deflection of a beam. The basic relationship that ties these concepts together is the relation between radius of curvature and the bending moment [Eq. (5-2)], which was developed in Chap. 5. Rewriting this equation for reference purposes, we have

$$\frac{1}{\rho} = \frac{M}{EI} \tag{7-6}$$

We now have all of the necessary beam relations before us, and so it is only necessary to organize them in an easily remembered group.

Putting Eqs. (7-6) and (7-3) together, we have

$$\frac{M}{EI} = \frac{d^2y}{dx^2} \tag{a}$$

If we differentiate Eq. (a) successively, twice, using Eqs. (7-4) and (7-5), we get

$$\frac{d}{dx}\left(\frac{M}{EI}\right) = \frac{V}{EI} = \frac{d^3y}{dx^3} \tag{b}$$

$$\frac{d}{dx}\left(\frac{V}{EI}\right) = \frac{q}{EI} = \frac{d^4y}{dx^4} \tag{c}$$

Displaying all these relations in a single group, we have

Loading: $\qquad\qquad\qquad \dfrac{q}{EI} = \dfrac{d^4y}{dx^4}$ $\qquad\qquad$ (7-7)

Shear force: $\qquad\qquad \dfrac{V}{EI} = \dfrac{d^3y}{dx^3}$ $\qquad\qquad$ (7-8)

Bending moment: $\qquad\quad \dfrac{M}{EI} = \dfrac{d^2y}{dx^2}$ $\qquad\qquad$ (7-9)

Slope:
$$\theta = \frac{dy}{dx}$$
(7-10)

Deflection:
$$y = f(x)$$
(7-11)

In the next several sections, we shall show how these important equations are used to determine beam deflections.

7-4. A UNIFORMLY LOADED BEAM   Though we shall seldom find it necessary to use the entire set of beam relations, their use can be demonstrated most simply by applying them to a uniformly loaded beam. And though the shear and moment relations were investigated in Chap. 4, we shall repeat the analysis here for the sake of continuity.

The entire set of beam relations, beginning with the loading diagram and ending with the deflection diagram, is displayed in Fig. 7-2. First, observe that all quantities, loading, shear, moment, slope, and deflection are positive if mea-

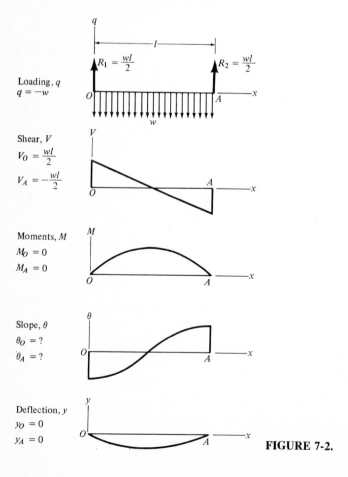

Loading, q
$q = -w$

Shear, V
$V_O = \dfrac{wl}{2}$
$V_A = -\dfrac{wl}{2}$

Moments, M
$M_O = 0$
$M_A = 0$

Slope, θ
$\theta_O = ?$
$\theta_A = ?$

Deflection, y
$y_O = 0$
$y_A = 0$

FIGURE 7-2.

sured upward, and negative if measured downward. Note, for example, that the loading $q = -w$ because w acts downward.

Also observe that the values of the various quantities are specified, when known, at the beam ends. Thus, the shear forces at the ends are the same, in magnitude, as the support reactions. And the bending moments and deflections at each end are zero, because of the supports.

Note that the slope θ is negative at O and positive at A. This is because a tangent to the deflection curve would slope downward at O (negative) and upward at A (positive). A tangent to the deflection curve at the middle of the beam ($x = l/2$) would be horizontal and indicate a zero slope, as shown on the slope diagram. Because of symmetry, we know that the maximum deflection occurs at the middle, and hence that the slope of the deflection curve is zero at the middle of the beam. On the other hand, the slope of the deflection curve at the ends is unknown, as shown. But, because of symmetry, the maximum deflection occurs at the middle, and so we do know that the slope of the beam is zero at the middle.

There are a rather large number of methods of solving the beam relations to be found in the literature. Widely known among these are the area-moment method, the conjugate-beam method, the use of singularity functions, and several energy methods. Because of the basic approach of this book, we shall employ the classical approach of simply integrating Eq. (7-7) successively until we eventually have found the deflection y. If we start with the loading, this requires four integrations. Frequently, it is possible to begin with the moment, in which case only two integrations are needed. The *double-integration* method, hence, is a well-known approach. There are three important advantages in the use of the classical approach:

1 The integrations are simple and require only a beginning knowledge of calculus.
2 The classic approach is the best way to define the problem if computers are to be employed using preprogrammed integration routines.
3 Graphical methods, illustrated in Chap. 4, are directly applicable.

To begin, substitute $q = -w$ into Eq. (7-7) to get

$$\frac{q}{EI} = \frac{d^4y}{dx^4} = -\frac{w}{EI} \tag{a}$$

This is the equation that we are to integrate four times. In the work below, we use the formulas for integration in Appendix A-6 and perform the integrations directly without the use of integral signs. For each integration, imagine that both sides of the equation are multiplied by dx and then integrated with respect to x. Starting with Eq. (a), here are the results:

$$\frac{V}{EI} = \frac{d^3y}{dx^3} = -\frac{wx}{EI} + C_1 \tag{b}$$

$$\frac{M}{EI} = \frac{d^2y}{dx^2} = -\frac{wx^2}{2EI} + C_1x + C_2 \tag{c}$$

$$\theta = \frac{dy}{dx} = -\frac{wx^3}{6EI} + \frac{C_1 x^2}{2} + C_2 x + C_3 \qquad (d)$$

$$y = f(x) = -\frac{wx^4}{24EI} + \frac{C_1 x^3}{6} + \frac{C_2 x^2}{2} + C_3 x + C_4 \qquad (e)$$

There are four constants of integration because four integrations were performed. These constants are obtained by substituting known values of the variables. The coordinate x is restricted to values lying between $x = 0$ and $x = l$. Thus, $x = 0$ and $x = l$ are the boundaries of x, and the corresponding values of V, M, θ, and y are called boundary values. The constants C are evaluated using known values of these boundary conditions. In particular, C_1 and C_2 are evaluated using known values of V and M, while C_3 and C_4 are obtained from known values of θ and y.

Let us use Eq. (c) to find C_1 and C_2. From Fig. 7-2, $M = 0$ at $x = 0$; we substitute these boundary conditions into Eq. (c) as follows:

$$0 = -\frac{w(0)}{2EI} + C_1(0) + C_2$$

Solving gives $C_2 = 0$. Next substitute the second boundary condition, $M = 0$ at $x = l$. This gives

$$0 = -\frac{wl^2}{2EI} + C_1 l$$

Whence, $C_1 = \dfrac{wl}{2EI}$

An alternative approach is to use Eq. (b) with the boundary conditions $V = wl/2$ at $x = 0$. This gives

$$\frac{wl}{2EI} = -\frac{w(0)}{EI} + C_1$$

or $C_1 = wl/2EI$, as before.

Before solving for C_3 and C_4, substitute the values of C_1 and C_2, just found, into Eqs. (d) and (e). The equations then become

$$\theta = -\frac{wx^3}{6EI} + \frac{wlx^2}{4EI} + C_3 \qquad (f)$$

$$y = -\frac{wx^4}{24EI} + \frac{wlx^3}{12EI} + C_3 x + C_4 \qquad (g)$$

As shown in Fig. 7-2, the slope is unknown at the boundaries. Therefore, we must utilize the knowledge that the deflection y is zero at both ends. Using Eq. (g) with $x = 0$ and $y = 0$ gives

$$C_4 = 0$$

But using $x = l$ and $y = 0$, we get

$$0 = -\frac{wl^4}{24EI} + \frac{wl^4}{12EI} + C_3 l$$

from which

$$C_3 = -\frac{wl^3}{24EI}$$

In this case, we can also obtain C_3 by noting from Fig. 7-2 that the slope is zero at the middle of the beam. This occurs because the beam is symmetrical. So, by using Eq. (f) and the knowledge that $\theta = 0$ at $x = l/2$, we could have obtained the same result for C_3.

Now that we have obtained the constants of integration, these can be substituted back into Eqs. (b) to (e) to get the equations for shear force, moment, slope, and deflection. The final results are

$$V = \frac{w}{2}(l - 2x) \tag{h}$$

$$M = \frac{wx}{2}(l - x) \tag{i}$$

$$\theta = \frac{w}{24EI}(6lx^2 - 4x^3 - l^3) \tag{j}$$

$$y = \frac{wx}{24EI}(2lx^2 - x^3 - l^3) \tag{k}$$

The maximum deflection occurs at the middle; by substituting $x = l/2$ into Eq. (k), we find it to be

$$y_{max} = -\frac{5wl^4}{384EI} \tag{l}$$

Finally, the slopes at $x = 0$ and $x = l$ are found from Eq. (j) to be

$$\theta_0 = -\frac{wl^3}{24EI} \qquad \theta_A = \frac{wl^3}{24EI} \tag{m}$$

7-5. A COMPUTER SOLUTION   Beam-deflection problems are rather easy to solve on the digital computer if an integration subroutine is available. Figure 7-3 is a block diagram showing the procedure. Each block in the diagram represents an integration subroutine, and so these blocks can be called integrators. Each integrator requires two inputs – the function to be integrated and the initial or starting value of the output. For example, the first integrator in Fig. 7-3 integrates the function $-w/EI$ to get V/EI, and the initial value of V is V_0.

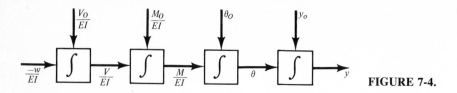

FIGURE 7-4.

The first step is to program the equations, by substituting known values of the loading and the geometry. If the example of Sec. 7-4 is being solved, then the starting value of the slope is unknown. So the computer is first requested to print out values of y at $x = l$ for various values of θ_0. The operator then repeats the trials until he finds a value of θ_0 for which $y = 0$ at $x = l$. With θ_0 known, the computer can be directed to solve the entire problem from $x = 0$ to $x = l$, and to print the output of all four integrators.

7-6. BEAM WITH CONCENTRATED LOAD   Figure 7-4 shows the loading, bending-moment, and shear-force diagrams for a simply supported beam with a concentrated loading not in the middle. The various boundary values shown have been obtained using the methods of Chap. 4. We are going to find the deflection by beginning with the bending moment, and hence we will integrate Eq. (7-9) twice.

First, we require the equation of M in terms of the loading and the geometry of the beam. As shown in Fig. 7-4, the moment is discontinuous at A. This

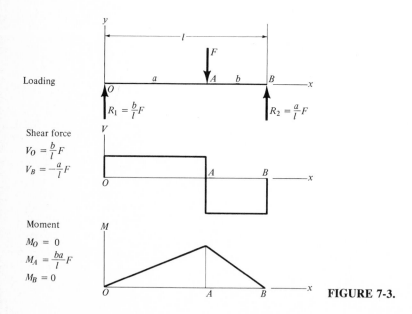

FIGURE 7-3.

means that we must write one equation for M between O and A and another between A and B. Using the methods of Chap. 4, we find these to be

$$M_{OA} = \frac{Fbx}{l} \qquad M_{AB} = \frac{Fbx}{l} - F(x-a) \qquad (a)$$

If we now apply Eq. (7-9) to OA, we have

$$\frac{M_{OA}}{EI} = \frac{d^2y}{dx^2} = \frac{Fbx}{EIl} \qquad (b)$$

Integrating successively gives

$$\theta_{OA} = \frac{dy}{dx} = \frac{Fbx^2}{2EIl} + C_1 \qquad (c)$$

$$y_{OA} = f(x) = \frac{Fbx^3}{6EIl} + C_1 x + C_2 \qquad (d)$$

Next, we apply the same technique to the portion of the beam between A and B. Equation (7-9) then may be written

$$\frac{M_{AB}}{EI} = \frac{d^2y}{dx^2} = \frac{Fbx}{EIl} - \frac{F(x-a)}{EI} \qquad (e)$$

Integrating twice yields

$$\theta_{AB} = \frac{dy}{dx} = \frac{Fbx^2}{2EIl} - \frac{F(x-a)^2}{2EI} + C_3 \qquad (f)$$

$$y_{AB} = f(x) = \frac{Fbx^3}{6EIl} - \frac{F(x-a)^3}{6EI} + C_3 x + C_4 \qquad (g)$$

In evaluating these four constants of integration which appear, we may use the conditions at the ends of the beam, as before. However, point A on the beam also constitutes a boundary. Thus, at A, the slope and deflection must have the same values no matter whether computed from Eqs. (c) and (d) or from Eqs. (f) and (g). This means that

$$\theta_{OA} = \theta_{AB} \qquad \text{and} \qquad y_{OA} = y_{AB} \text{ at } x = a$$

Since the slopes are unknown at the ends, we have left only two usable end conditions

$$y_{OA} = 0 \text{ at } x = 0 \qquad \text{and} \qquad y_{AB} = 0 \text{ at } x = l$$

These four conditions are sufficient. Taking Eq. (d) first, we place $y_{OA} = 0$ and $x = 0$ to get $C_2 = 0$. Next, at $x = a$, $\theta_{OA} = \theta_{AB}$. Therefore, place Eqs. (c) and (f) equal to each other. This gives

$$\frac{Fba^2}{2EIl} + C_1 = \frac{Fba^2}{2EIl} - 0 + C_3$$

Thus, $C_1 = C_3$. We have two conditions left. At $x = a$, $y_{OA} = y_{AB}$. Therefore, place Eq. (d) equal to (g), using $C_2 = 0$ and $C_3 = C_1$. The result is

$$\frac{Fba^3}{6EIl} + C_3 a = \frac{Fba^3}{6EIl} - 0 + C_3 a + C_4$$

Since all other terms cancel, $C_4 = 0$. The last condition is that $y_{AB} = 0$ at $x = l$. Placing these conditions in Eq. (g) yields

$$0 = \frac{Fbl^3}{6EIl} - \frac{F(l-a)^3}{6EI} + C_3 l$$

Substituting b for $l - a$ and solving for C_3 gives

$$C_3 = C_1 = \frac{Fb}{6EIl}(b^2 - l^2)$$

If we now place the constants into the equations, we finally obtain for the deflection

$$y_{OA} = \frac{Fbx}{6EIl}(x^2 + b^2 - l^2) \tag{h}$$

$$y_{AB} = \frac{Fb}{6EIl}\left[x^3 - \frac{l}{b}(x-a)^3 + (b^2 - l^2)x\right] \tag{i}$$

The maximum deflection occurs at a value of x for which the slope of the deflection curve is zero. If we have a beam with $a > b$, this will occur between points O and A. By substituting C_1 into Eq. (c), the slope equation between O and A is found to be

$$\theta_{OA} = \frac{Fbx^2}{2EIl} + \frac{Fb}{6EIl}(b^2 - l^2)$$

$$= \frac{Fbx^2}{2EIl} - \frac{Fb}{6EIl}(l^2 - b^2)$$

$$= \frac{Fb}{2EIl}\left(x^2 - \frac{l^2 - b^2}{3}\right)$$

Thus, making $\theta_{OA} = 0$, we obtain

$$\frac{Fb}{2EIl}\left(x^2 - \frac{l^2 - b^2}{3}\right) = 0$$

Solving this equation for x gives the location of maximum deflection at

$$x = \sqrt{\frac{l^2 - b^2}{3}} \tag{j}$$

Upon substituting this value of x into Eq. (h), we find, after some algebraic manipulation, that the maximum deflection is

$$y_{max} = -\frac{Fb(l^2 - b^2)^{3/2}}{9\sqrt{3}\,EIl} \tag{k}$$

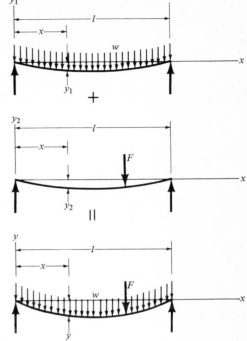

FIGURE 7-5.

An interesting conclusion, rather useful for design purposes, arises from a closer examination of Eq. (j). If we place the force at the middle of the beam, then Eq. (j) gives $x = l/2$ for the location of the maximum deflection, as expected. On the other hand, as we move the load toward the support farther from the middle, b gets smaller. In the limit, b nears zero and $x = l/\sqrt{3} = 0.577l$. Thus, regardless of the position of the load, the maximum deflection is always near the middle of the beam. So, for a quick estimate of y_{max}, use $x = l/2$ in the deflection formula.

7-7. SUPERPOSITION   When a beam has more than one load or type of loading, then the deflection can be obtained by adding the deflections that would be produced by each load acting separately. This is called the *method of superposition*, and it is illustrated in Fig. 7-5. Here, for example, the total deflection y, at any point x, is obtained by adding the deflection y_1 of the uniformly loaded beam to the deflection y_2 caused by the concentrated load.

7-8. DEFLECTIONS BY GRAPHICAL INTEGRATION   In Sec. 4-15, the method of graphical integration was used to derive the shear-force and bending-moment diagrams from the loading diagram. Beginning with the bending-moment diagram, two more such integrations will yield the slope and deflection diagrams.

When this method is used, we usually begin with the M/EI diagram rather than the M diagram. One advantage of the method of graphical integration is that the beam need not have the same moment of inertia throughout its length. Because of the fact that the slope at the ends of simply supported beams is unknown, we shall illustrate the method with the following example. The scales used are defined in Sec. 4-15.

EXAMPLE 7-1   The problem selected for demonstration of the method is illustrated in Fig. 7-6a. This is a shouldered steel shaft, commonly found in rotating machinery. The shoulders are used to locate gears, bearings, pulleys, and the like in an axial direction, and to resist thrust loads when they are present. In this case, the effect of the bearings and gears, say, have been replaced by the forces that they exert on the shaft. This means, as shown, that the beam is assumed to be simply supported and that the distributed forces exerted by the parts on the shaft have been replaced by their resultants. The bending moment diagram is shown in Fig. 7-6b. Find the deflection diagram, its scale, and the maximum deflection with its location.

SOLUTION

The first step is to compute the M/EI values corresponding to every point of discontinuity on the diagram which results. The computation is most convenient if made in tabular form. This has been done, and the results are displayed in Table 7-1. The moment of inertia is computed from the equation $I = \pi d^4/64$. Then, M/EI is computed using $E = 30(10)^6$ psi.

In Fig. 7-7, the shaft drawing is reproduced for reference purposes using the scale $S_x = 4$ in/in. Note that these are two kinds of inches, and so they cannot be canceled. Thus, 4 in/in means every 4 in of shaft is represented by a 1 in space on

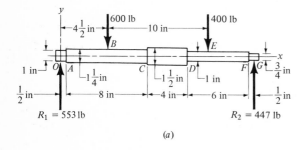

(a)

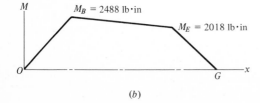

(b)

FIGURE 7-6.

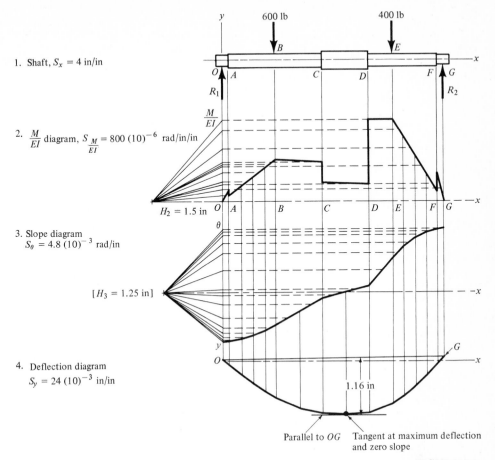

1. Shaft, $S_x = 4$ in/in

2. $\frac{M}{EI}$ diagram, $S_{\frac{M}{EI}} = 800\,(10)^{-6}$ rad/in/in

3. Slope diagram
   $S_\theta = 4.8\,(10)^{-3}$ rad/in

   $[H_3 = 1.25$ in$]$

4. Deflection diagram
   $S_y = 24\,(10)^{-3}$ in/in

$H_2 = 1.5$ in

Parallel to OG Tangent at maximum deflection
and zero slope

1.16 in

FIGURE 7-7.

the drawing. The M/EI diagram is now plotted in line with the shaft drawing using any convenient scale, but, of course, using the same scale for x. In this case, a scale $S_{M/EI} = 800(10)^{-6}$ rad/in/in means that an $M/EI = 1437(10)^{-6}$ rad/in at D occupies about 1.8 in of drawing, which gives a moderate sized diagram.

Before integrating, we divide the diagram into vertical strips, using more strips for the steep portions in order to get a smoother integral. Now choose the pole distance H_2 and proceed with the graphical integration as described in Chap. 4. The result is the slope diagram. The scale computed is

$$S_\theta = H_2 S_{M/EI} S_x = (1.5)\,(800)\,(10)^{-6}(4) = 4800(10)^{-6} \text{ rad/in}$$

or $4.8(10)^{-3}$ rad/in.

Before integrating the slope diagram, choose a place to draw the x axis. Usually, this will be chosen so as to cross the slope diagram near the middle of the beam. This does not mean that the slope is zero at this crossing at all, but only that we think the slope is close to zero at the slope-curve crossing. In other words, we do not know where this crossing should be, so we guess at it.

TABLE 7-1

shaft segment	moment, lb · in	moment of inertia, in⁴	M/EI, rad/in
OA	$M_O = 0$ $M_A = 276$	0.049	0 $188(10)^{-6}$
AB	$M_A = 276$ $M_B = 2488$	0.120	$76.7(10)^{-6}$ $691(10)^{-6}$
BC	$M_B = 2488$ $M_C = 2300$	0.120	$691(10)^{-6}$ $639(10)^{-6}$
CD	$M_C = 2300$ $M_D = 2112$	0.248	$309(10)^{-6}$ $284(10)^{-6}$
DE	$M_D = 2112$ $M_E = 2018$	0.049	$1437(10)^{-6}$ $1373(10)^{-6}$
EF	$M_E = 2018$ $M_F = 224$	0.049	$1373(10)^{-6}$ $152(10)^{-6}$
FG	$M_F = 224$ $M_G = 0$	0.0156	$479(10)^{-6}$ 0

Next, choose a pole distance H_3 and integrate again. This yields the deflection diagram as shown. Note that more vertical strips were used to obtain a smoother integral. The closing chord, OG in this case, will seldom be a horizontal line. The location of the point of maximum deflection, and hence of zero slope, is found by drawing a tangent to the deflection curve parallel to the closing chord. In this case, the scale of the deflection diagram is

$$S_y = H_3 S_\theta S_x = (1.25)(4.8)(10)^{-3}(4) = 24(10)^{-3} \text{ in/in}$$

To get the maximum deflection, measure the *vertical* distance between the tangent and the closing chord. This is 1.16 in of drawing. Therefore, the maximum deflection is

$$y_{max} = 1.16 S_y = 1.16(24)(10)^{-3} = 0.0278 \text{ in} \qquad Ans.$$

This is located, as shown, at $x = 10.5$ in. ////

PROBLEMS[1]

SECTIONS 7-4 TO 7-6

7-1 to 7-10   Verify the solutions to beams No. 1 through No. 10, respectively, in Table A-20.

SECTION 7-7

7-11   **(a)** Find the deflection of point *B* of the cantilever shown in the figure. Note that the two structural steel angles are back-to-back.

   (b) As shown, a single angle supports the two loads. Find the deflection of the beam at midspan.

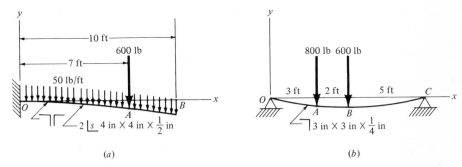

(a)   (b)

PROB. 7-11.

7-12   **(a)** Find the deflection of the steel shaft at point *A*.

   (b) A rectangular steel bar supports the two overhanging loads. Find the deflection at the ends and in the middle. (*Hint:* use Table A-20-9.)

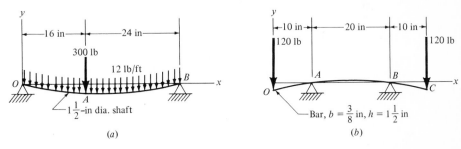

(a)   (b)

PROB. 7-12.

7-13   **(a)** A structural steel I beam supports the two concentrated loads and a uniform external load of 100 lb/ft. By including the weight of the beam in the distributed loading, find the maximum deflection.

   (b) Two concentrated loads are supported by the 6-in structural steel channel. Find the deflection at the end of the cantilever.

---

[1] Unless specified, neglect the weight of the beam in deflection analysis.

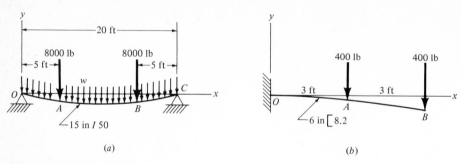

(a)   (b)

PROB. 7-13.

7-14 (a) A 4-in structural-steel channel supports the distributed load and the concentrated load, as shown. Find the deflection at point *A*.
      (b) Find the deflection at the middle of the 2-in diameter steel shaft.

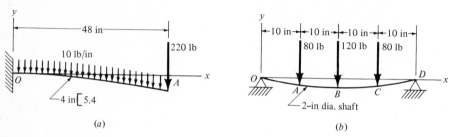

(a)   (b)

PROB. 7-14.

7-15 The figure shows a cantilever steel spring which is to have a rectangular cross section of proportions $b = 8h$. Determine the dimensions of the spring such that it has a scale $k = 140$ lb/in. [see Eq. (6-16)]. What is the maximum bending stress if the operating range is $\frac{1}{4}$ in?

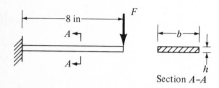

Section *A–A*

PROB. 7-15.

7-16 Illustrated is a rectangular steel bar with simple supports at the ends and loaded by a force *F* at the middle, which is to be designed as a spring. The ratio of the width to the thickness is to be $b = 16h$, and the desired spring scale is 2400 lb/in.
      (a) Find the cross-section dimensions.
      (b) What is the maximum permissible spring deflection if the yield strength of the material is 90 kpsi?

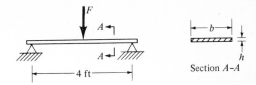

PROB. 7-16.

7-17   A 4-ft steel ruler is to be manufactured from a bar $\frac{3}{16}$ in thick. The depth of the section is to be such that when the ruler is balanced exactly in the center, the overhanging ends will deflect less than 0.001 in. What should be the section depth? Does the thickness matter?

7-18   Illustrated in the figure is a $1\frac{1}{2}$-in diameter steel countershaft which supports two pulleys. Pulley *A* delivers power to a machine causing a tension of 600 lb in the tight side of the belt and 80 lb in the loose side, as indicated. Pulley *B* receives power from a motor. The belt tensions on pulley *B* have the relation $T_1 = 0.125T_2$. Find the deflection of the shaft in the *z* direction at pulleys *A* and *B*. Assume that the bearings constitute simple supports.

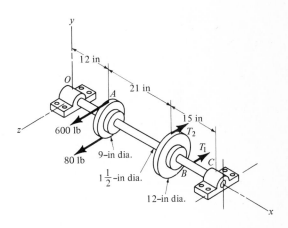

PROB. 7-18.

7-19   The figure shows a steel countershaft which supports two pulleys. Pulley *C* receives power from a motor producing the belt tensions shown. Pulley *A* transmits this power to another machine through the belt tensions T_1 and T_2 such that $T_1 = T_2/8$. Find the deflection of the overhanging end of the shaft. Assume simple supports for the bearings.

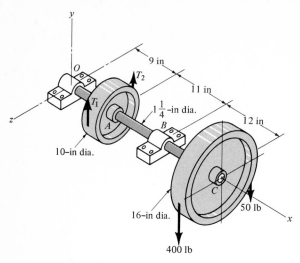

PROB. 7-19.

SECTION 7-8

    7-20 The figure shows the drawing of a steel shaft and its loading diagram. Both bearings are self-aligning and so the shaft is simply supported. Using graphical integration, find the maximum deflection.

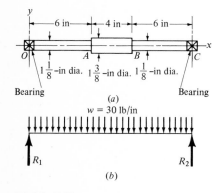

Bearing   Bearing

(a)

PROB. 7-20.

    7-21 Determine the maximum deflection of the simply supported steel shaft shown in the figure. Use graphical integration.

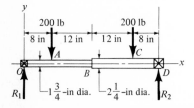

PROB. 7-21.

SI UNITS

7-22 A short round cantilever steel spring is to be 240 mm in length and support a concentrated bending force F at the end. Determine the diameter of the bar if it is to have a spring scale of 35 kN/m. What is the maximum bending stress in the spring if the operating range is 8 mm?

7-23 The cantilever steel spring shown for Prob. 7-15 is to have a length of 4 m instead of the length shown. The cross section is to be proportioned so that $b = 5h$. Determine the dimensions of the spring such that it has a scale of 400 kN/m. What maximum stress would be produced by a deflection of 5 mm at the free end?

statically
indeterminate problems

8-1. AXIALLY LOADED MEMBERS   A problem in which the laws of statics are not sufficient to determine all the unknown forces or moments is said to be *statically indeterminate*. Such problems are solved by writing the appropriate equations of static equilibrium and additional equations pertaining to the deformation of the part. In all, the number of equations must equal the number of unknowns.

For axially loaded members, Eq. (2-7) is used to obtain the deformation equations. It is

$$\delta = \frac{Fl}{AE} \tag{8-1}$$

But sometimes it is convenient to think of axially loaded members as springs. The spring rate or spring constant, from Eq. (6-17), may be written as

$$k = \frac{F}{\delta} = \frac{AE}{l} \tag{8-2}$$

and so the deformation may also be expressed as

$$\delta = \frac{F}{k} \tag{8-3}$$

A simple example of a statically indeterminate problem is furnished by the nested helical springs in Fig. 8-1*a*. When this assembly is loaded by the compressive force F, it deforms through the distance δ. What is the compressive force in each spring?

Only one equation of static equilibrium can be written. It is

$$\Sigma F = F - F_1 - F_2 = 0 \tag{a}$$

which simply says that the total force F is resisted by a force F_1 in spring 1 plus the force F_2 in spring 2. Since there are two unknowns, F_1 and F_2, Eq. (*a*) cannot be solved; thus, the system is statically indeterminate.

To write another equation, note the deformation relation in Fig. 8-1*b*. The two springs have the same deformation. Thus, we obtain the second equation as

$$\delta_1 = \delta_2 = \delta \tag{b}$$

If we now substitute Eq. (8-3) in (*b*), we have

$$\frac{F_1}{k_1} = \frac{F_2}{k_2} \tag{c}$$

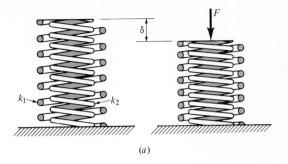

(a)

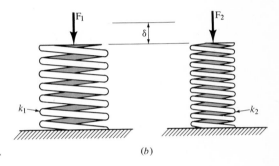

FIGURE 8-1.   (b)

Now solve Eq. (c) for F_1 and substitute the result in Eq. (a). This gives

$$F - \frac{k_1}{k_2} F_2 - F_2 = 0$$

or $F_2 = \dfrac{k_2 F}{k_1 + k_2}$ (d)

This completes the solution because, with F_2 known, F_1 can be found from Eq. (c).

EXAMPLE 8-1   A cable is made using a 16 ga (0.0625 in) steel wire and three strands of 12 ga (0.08081 in) copper wire. Find the stress in each wire if the cable is subjected to a tension of 250 lb.

SOLUTION

Let the tension in each copper wire be F_c and the tension in the steel wire be F_s. Then, for static equilibrium,

$$\Sigma F = 250 - 3F_c - F_s = 0$$ (1)

To get the second equation, we observe that the copper wires elongate the same amount as the steel wire. Therefore, continuing the use of the subscripts c for copper and s for steel,

$$\delta_c = \delta_s$$ (2)

Substituting Eq. (8-1) into (2) gives

$$\frac{F_c l}{A_c E_c} = \frac{F_s l}{A_s E_s}$$

Canceling the length l and solving for F_c gives

$$F_c = \frac{A_c E_c}{A_s E_s} F_s \tag{3}$$

From Table A-9, we find $E_c = 17.2$ Mpsi and $E_s = 30$ Mpsi. The areas are

$$A_c = \frac{\pi d^2}{4} = \frac{\pi (0.08081)^2}{4} = 5.13(10)^{-3} \text{ in}^2$$

$$A_s = \frac{\pi (0.0625)^2}{4} = 3.07(10)^{-3} \text{ in}^2$$

Substituting these values into Eq. (3) yields

$$F_c = \frac{[5.13(10)^{-3}][17.2(10)^6]}{[3.07(10)^{-3}][30(10)^6]} F_s = 0.958 F_s \tag{4}$$

Next we substitute this relation for F_c into Eq. (1) and solve the result for F_s. We then get

$$F_s = \frac{250}{3(0.958) + 1} = 64.5 \text{ lb}$$

And Eq. (4) yields

$$F_c = 0.958(64.5) = 61.8 \text{ lb}$$

Finally, we find the tensile stresses in the copper and in the steel to be

$$\sigma_c = \frac{F_c}{A_c} = \frac{61.8}{5.13(10)^{-3}} = 12\ 000 \text{ psi} \qquad Ans.$$

$$\sigma_s = \frac{F_s}{A_s} = \frac{64.5}{3.07(10)^{-3}} = 21\ 000 \text{ psi} \qquad Ans. \qquad ////$$

EXAMPLE 8-2   A built-up connecting rod consists of three steel bars, each $\frac{1}{4}$ in thick and $1\frac{1}{4}$ in wide, as shown in Fig. 8-2. During assembly it was discovered that one of the bars measured only 31.997 in between pin centers, the other two bars measuring exactly 32.000 in in length.

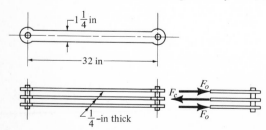

FIGURE 8-2.

(a) Determine the stress in each bar after assembly, but before an external load is applied.

(b) Determine the stress in each bar after an external tensile load of 5000 lb is applied to the assembled connecting rod. Assume a uniform cross section between pin centers.

SOLUTION

(a) Let us assume that the short bar is placed in the middle. Then, if we assign the subscript c to the center bar and the subscript o to the outer bars, static equilibrium requires that

$$\Sigma F = F_c - 2F_o = 0 \tag{1}$$

because the external force is zero. There are two unknowns in this equation, and so it cannot be solved. When the bars are assembled, the center bar stretches and the outer bars compress. If, say, the center bar stretches 0.001 in, then the outer bars must be compressed 0.002 in to get the pins assembled. In other words, the elongation of the center bar plus the compression of the outer bars must equal 0.003 in. In equation form,

$$\delta_c + \delta_o = 0.003 \tag{2}$$

where δ_c = extension of center bar, in
δ_o = compression of outer bars, in

We can solve Eqs. (1) and (2) simultaneously for F_c and F_o by the use of Eq. (8-1). Equation (2) becomes

$$\frac{F_c l}{AE} + \frac{F_o l}{AE} = 0.003 \tag{3}$$

Introducing the fact that $F_c = 2F_o$ into Eq. (3) yields

$$\frac{2F_o l}{AE} + \frac{F_o l}{AE} = 0.003$$

from whence

$$F_o = \frac{0.003AE}{3l} = \frac{(0.003)[(0.25)(1.25)](30)(10)^6}{3(32)} = 293 \text{ lb}$$

Then,

$$F_c = 2F_o = (2)(293) = 586 \text{ lb}$$

Thus, the stresses are

$$\sigma_c = \frac{F_c}{A} = \frac{586}{(0.25)(1.25)} = 1880 \text{ psi} \qquad Ans.$$

$$\sigma_o = \frac{F_o}{A} = \frac{293}{(0.25)(1.25)} = -940 \text{ psi} \qquad Ans.$$

(b) When an external tensile load is applied, the condition of static equilibrium is

$$\Sigma F = 5000 - F_c - 2F_o = 0 \tag{4}$$

For the deformation relation, we observe that the center bar will stretch an amount equal to 0.003 in plus the stretch of the outer bars. In other words,

$$\delta_c = \delta_o + 0.003 \tag{5}$$

Introducing Eq. (8-1) into (5) again, gives

$$\frac{F_c l}{AE} = \frac{F_o l}{AE} + 0.003 \tag{6}$$

Using $A = (0.25)(1.25) = 0.312$ in², $l = 32$ in, and $E = 30$ Mpsi, Eqs. (4) and (6) are solved simultaneously. The results are

$$F_o = 1374 \text{ lb} \qquad F_c = 2252 \text{ lb}$$

Then, as before,

$$\sigma_c = \frac{F_c}{A} = \frac{2252}{0.312} = 7200 \text{ psi} \qquad Ans.$$

$$\sigma_o = \frac{F_o}{A} = \frac{1374}{0.312} = 4400 \text{ psi} \qquad Ans. \qquad ////$$

8-2. BOLTED JOINTS LOADED IN TENSION

Figure 8-3 shows a pressure vessel that consists of a double-flanged cylinder with a cap or cylinder head bolted to each end. The bolts extend the full length of the cylinder in this case, although alternate fastening methods can be devised. When the nuts are tightened, usually with a torque or an impact wrench, the bolts are placed in tension. In fact, if too much wrench torque is applied, the bolts may fail due to a combination of high bolt tension and some bolt torque caused by friction in the threads.

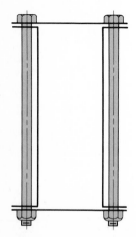

FIGURE 8-3.

When the nuts are properly tightened, a rather large amount of bolt tension exists. The reaction to this tension is the compression of the seal material, if any, and of the cylinder walls. Now when a fluid under pressure is introduced into the cylinder, its effect is to reduce the compression in the cylinder walls and to increase the tension in the bolts. The purposes of this section include the determination of the amount of this stress change. Since the problem is statically indeterminate, it is a practical example of the use of the material of the previous section. As indicated in the following example, the problem can be studied without introducing new information.

EXAMPLE 8-3 Figure 8-4 shows a pressure cylinder 4 in in diameter which uses 6 SAE grade 5 bolts (see Table 3-1) having a grip of 12 in. These bolts are to be tightened to 90 percent of the proof strength in accordance with the recommendation in Sec. 3-6.

(a) Find the tensile stress in the bolts and the compressive stress in the cylinder walls.

(b) Find the tensile stress in the bolts and the compressive stress in the cylinder walls after a fluid under a pressure of 600 psi is introduced into the cylinder.[1]

SOLUTION

(a) From Table 3-1, we find the proof strength of the bolts to be 85 kpsi. Therefore, each bolt is to be tightened to produce a tensile stress of

$$\sigma_b = 0.9S_P = 0.9(85)(10)^3 = 76.5(10)^3 \text{ psi} \qquad Ans.$$

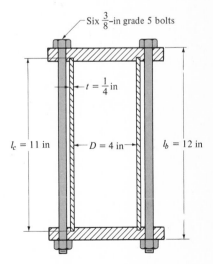

FIGURE 8-4.

---

[1] Pressure vessels should always be designed using an approved safety code. This problem is intended to be used to demonstrate an important and useful effect and probably does not conform. For proper design, see the ASME Boiler and Pressure Vessel Code, available from the American Society of Mechanical Engineers.

The cross-sectional area of a bolt is

$$A_b = \frac{\pi d^2}{4} = \frac{\pi (0.375)^2}{4} = 0.110 \text{ in}^2$$

Since there are six bolts, the total preload is

$$F_i = 6\sigma_b A_b = (6)(76.5)(10)^3(0.110) = 50.5(10)^3 \text{ lb}$$

A close and easy approximation to the cross-sectional area of the cylinder is obtained by multiplying the circumference by the thickness. This gives

$$A_c = \pi Dt = \pi (4)(0.25) = 3.14 \text{ in}^2$$

Thus, the compressive stress in the cylinder walls is

$$\sigma_c = \frac{F_i}{A_c} = -\frac{50.5(10)^3}{3.14} = -16.1(10)^3 \text{ psi} \qquad Ans.$$

(b) A pressure of 600 psi acting against the two cylinder heads will produce a force

$$P = \frac{\pi D^2}{4} p = \frac{\pi (4)^2}{4}(600) = 7540 \text{ lb}$$

This force will increase the tensile stress in the bolts and reduce the compressive stress in the cylinder walls. Our problem is to determine the amount of this change. Let P_b be the portion of P taken by the bolts and P_c the portion which reduces the compression of the members. Then

$$P = P_b + P_c \tag{1}$$

Since the bolt tension increases by the amount P_b, the bolts will elongate. Designate this elongation as $\Delta\delta_b$. Also, the cylinder loses some of its compression because of P_c. Designate this loss of compression as $\Delta\delta_c$. Then

$$\Delta\delta_c = \Delta\delta_b \tag{2}$$

Equation (8-1) can be used to find both sides of Eq. (2). For the cylinder, we have

$$\Delta\delta_c = \frac{P_c l_c}{A_c E} = \frac{P_c(11)}{(3.14)(30)(10)^6} = 117(10)^{-3}P_c$$

Since there are six bolts,

$$\Delta\delta_b = \frac{P_b l_b}{6A_b E} = \frac{P_b(12)}{(6)(0.110)(30)(10)^6} = 606(10)^{-3}P_b$$

Substituting these into Eq. (2) and solving for P_c gives

$$P_c = \frac{606(10)^{-3}P_b}{117(10)^{-3}} = 5.18P_b$$

Since $P = 7540$ lb, Eq. (1) gives

$$7540 = P_b + 5.18P_b$$

or

$$P_b = \frac{7540}{6.18} = 1220 \text{ lb}$$

and

$$P_c = 5.18P_b = 5.18(1220) = 6320 \text{ lb}$$

In other words, the bolt tension is increased by 1220 lb or about 200 lb per bolt. And the members lose 6320 lb of the precompression. The resulting bolt load is

$$F_b = F_i + P_b = 50.5(10)^3 + 1220 = 51.7(10)^3 \text{ lb}$$

Therefore, the tensile stress in the bolts is

$$\sigma_b = \frac{F_b}{A_b} = \frac{51.7(10)^3}{(6)(0.110)} = 78.3(10)^3 \text{ psi} \qquad Ans.$$

Compare this with the prestress of $76.5(10)^3$ psi. The bolt tension has increased by 1800 psi.

The resultant compression in the cylinder walls is

$$F_c = F_i + P_c = -50.5(10)^3 + 6320 = -44.2(10)^3 \text{ lb}$$

And so the compressive stress now becomes

$$\sigma_c = \frac{F_c}{A_c} = \frac{-44.2(10)^3}{3.14} = -14.1(10)^3 \text{ psi} \qquad Ans.$$

which is a change of 2000 psi. The importance of this example, however, is that most of the external load, 6320 lb out of a total of 7540 lb, goes to reduce the compression of the cylinder rather than to increase the tension in the bolts. The limit, of course, would be reached when the external or pressure load removes all of the compression. In such a case, all of the external load would go into the bolts, probably causing a bolt failure. ////

8-3. TORSIONALLY LOADED MEMBERS

A statically indeterminate problem involving torsion is really no different than for axially loaded members. Torques or moments are involved instead of forces and the deformations are measured in angular units instead. For a round bar subject to torsion, the angular deflection in radians is

$$\theta = \frac{Tl}{GJ} \tag{8-4}$$

from Eq. (6-1).

An example is shown in Fig. 8-5. Here a core of diameter d_i is surrounded by a shell of a different material of diameter d_o. If a torque T is applied to this built-up bar, how much torque is carried by each part?

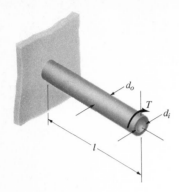

FIGURE 8-5.

The equilibrium equation is

$$\Sigma T = T - T_i - T_o = 0 \qquad (a)$$

Both parts are twisted through the same angle, and so, using Eq. (8-4), the deformation relation is

$$\frac{T_i l}{G J_i} = \frac{T_o l}{G J_o} \qquad (b)$$

where J_i and J_o are the polar area moments of inertia of the inner and outer parts, respectively. Equations (a) and (b) may now be solved simultaneously for the unknown torques.

8-4. BEAMS   An indeterminate beam is one in which there are more support reactions than can be found using the principles of static equilibrium. After the statics equations are written, the remaining reactions must be found using information concerning the slope or deflection of the beam. Basically, this is no different than our previous approach to solving problems that are statically indeterminate. It is a more complicated problem because there are two or more integrations that might be necessary in the process.

As a first example of the procedure, we select the beam of Table A-20-11, which has one end fixed and one simply supported. The beam has a concentrated load in the center. Figure 8-6a is a reproduction of the beam, and Fig. 8-6b is a free-body diagram. Applying static equilibrium, we have

$$\Sigma F = R_1 - F + R_2 = 0 \qquad (a)$$

$$\Sigma M_0 = M_1 - \frac{Fl}{2} + R_2 l = 0 \qquad (b)$$

Thus, we have two equations containing the three unknown reactions R_1, M_1, and R_2. The shear-force and bending-moment diagrams, in terms of these unknown reactions, are shown in Fig. 8-6c and d. The boundary values are

$$V_0 = R_1 \qquad V_B = -R_2 \qquad M_0 = -M_1 \qquad M_B = 0$$

These, of course, correspond to $x = 0$ and $x = l$. Note that the moment reaction at the fixed end is negative.

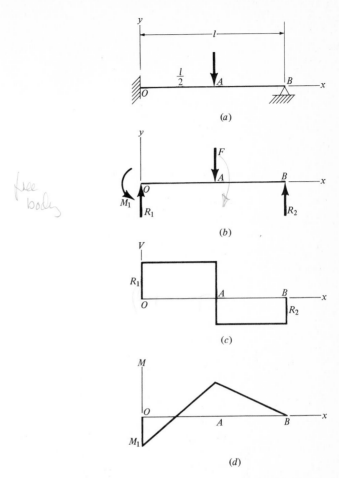

FIGURE 8-6.

Since the bending-moment diagram has a point of discontinuity at A, we must write two moment equations, one from O to A and another from A to B. Each of these moment equations must be integrated twice to get the corresponding deflection equations. Each integration yields a constant of integration. Altogether, therefore, there will be four constants of integration to evaluate. There are already three unknown reactions. So a total of seven unknowns are involved, and hence seven independent relations must be found to evaluate them all.

The bending moment from O to A is

$$M_{OA} = -M_1 + R_1 x$$

which is valid only for x's between 0 and $l/2$. Substituting into Eq. (7-9) gives

$$M_{OA} = EI \frac{d^2 y}{dx^2} = -M_1 + R_1 x \qquad (c)$$

Integrating twice gives the slope and deflection relations as

$$EI\theta_{OA} = EI \frac{dy}{dx} = -M_1 x + \frac{R_1 x^2}{2} + C_1 \tag{d}$$

$$EIy_{OA} = EIf(x) = -\frac{M_1 x^2}{2} + \frac{R_1 x^3}{6} + C_1 x + C_2 \tag{e}$$

We next repeat the same procedure for the section of beam from A to B. The moment is

$$M_{AB} = -M_1 + R_1 x - F\left(x - \frac{l}{2}\right) = -M_1 + R_1 x - Fx + \frac{Fl}{2}$$

which is valid only for x's between 0 and $l/2$. As before, we substitute into Eq. (7-9) to get

$$M_{AB} = EI \frac{d^2 y}{dx^2} = -M_1 + R_1 x - Fx + \frac{Fl}{2} \tag{f}$$

Again, integrating twice yields

$$EI\theta_{AB} = EI \frac{dy}{dx} = -M_1 x + \frac{R_1 x^2}{2} - \frac{Fx^2}{2} + \frac{Flx}{2} + C_3 \tag{g}$$

$$EIy_{AB} = EIf(x) = -\frac{M_1 x^2}{2} + \frac{R_1 x^3}{6} - \frac{Fx^3}{6} + \frac{Flx^2}{4} + C_3 x + C_4 \tag{h}$$

Equations (a) through (h) can now be used to find the four constants of integration and the three beam reactions. To simplify the presentation, we shall first describe the steps to be followed and present the results of these steps. Then the actual algebra involved will be shown separately in fine print.

1   At $x = 0$, the slope $\theta_{OA} = 0$ because the left end of the beam is fixed. Therefore, place $x = 0$ and $\theta_{OA} = 0$ in Eq. (d) and solve for C_1. This gives

$$C_1 = 0 \tag{i}$$

2   At $x = 0$, the deflection $y_{OA} = 0$ because this is at a beam support. So place $x = 0$ and $y_{OA} = 0$ in Eq. (e) and solve for C_2. The result is

$$C_2 = 0 \tag{j}$$

3   At point A on the beam, the slope given by Eq. (d) must be the same as the slope given by Eq. (g). Therefore, place Eqs. (d) and (g) equal to each other, substitute $x = l/2$ and $C_1 = 0$, and solve the result for C_3. This yields

$$C_3 = -\frac{Fl^2}{8} \tag{k}$$

4   At point A on the beam, the deflection given by Eq. (e) must be the same as the deflection given by Eq. (h). So place Eqs. (e) and (h) equal to each other, substitute $x = l/2$ and the known values of C_1, C_2, and C_3, and solve the result for C_4.

Thus, we get

$$C_4 = \frac{Fl^3}{48} \tag{l}$$

5   As indicated in Fig. 8-6d, $M_B = 0$. So, using Eq. (f), substitute $M_{AB} = 0$ and $x = l$. The result is

$$M_1 - R_1 l = -\frac{Fl}{2} \tag{m}$$

6   The deflection is zero at point B on the beam. Therefore, place $x = l$ and $y_{AB} = 0$ in Eq. (h). Also, substitute the known values of C_3 and C_4. Upon simplifying and rearranging, we have

$$M_1 - \frac{R_1 l}{3} = -\frac{Fl}{24} \tag{n}$$

7   Solve Eqs. (m) and (n) simultaneously for M_1 and R_1. The results are

$$M_1 = \frac{3Fl}{16} \qquad R_1 = \frac{11F}{16} \tag{o}$$

8   Using the known value of R_1, solve Eq. (a) for R_2. This gives

$$R_2 = \frac{5F}{16} \tag{p}$$

Here is the algebra involved in the 8 previous steps:

1   Using Eq. (d), we have

$$0 = -M_1(0) + \frac{R_1(0)^2}{2} + C_1 \qquad C_1 = 0$$

2   With Eq. (e)

$$0 = -\frac{M_1(0)^2}{2} + \frac{R_1(0)^3}{6} + C_1(0) + C_2 \qquad C_2 = 0$$

3   First place Eqs. (d) and (g) equal to each other, using $C_1 = 0$. This gives

$$-M_1 x + \frac{R_1 x^2}{2} = -M_1 x + \frac{R_1 x^2}{2} - \frac{Fx^2}{2} + \frac{Flx}{2} + C_3$$

Now cancel the first two terms on the right with both terms on the left, substitute $l/2$ for x, and solve the result for C_3. Thus, we have

$$0 = -\frac{Fl^2}{8} + \frac{Fl^2}{4} + C_3 \qquad C_3 = -\frac{Fl^2}{8}$$

4   First we place Eq. (e) equal to Eq. (h) using the known values of C_1, C_2, and C_3. This gives

$$-\frac{M_1 x^2}{2} + \frac{R_1 x^3}{6} = -\frac{M_1 x^2}{2} + \frac{R_1 x^3}{6} - \frac{Fx^3}{6} + \frac{Flx^2}{4} - \frac{Fl^2 x}{8} + C_4$$

Of course this equation is only valid for $x = l/2$. Now cancel the first two terms on the right with those on the left, substitute $x = l/2$, and solve for C_4.

$$0 = -\frac{Fl^3}{48} + \frac{Fl^3}{16} - \frac{Fl^3}{16} + C_4 \qquad C_4 = \frac{Fl^3}{48}$$

5 Substitute $M_{AB} = 0$ and $x = l$ in Eq. (f). This gives

$$0 = -M_1 + R_1 l - Fl + \frac{Fl}{2}$$

Rearranging, we get

$$M_1 - R_1 l = -\frac{Fl}{2}$$

6 First substitute the known values of C_3 and C_4 into Eq. (h). This gives

$$EIy_{AB} = -\frac{M_1 x^2}{2} + \frac{R_1 x^3}{6} - \frac{Fx^3}{6} + \frac{Flx^2}{4} - \frac{Fl^2 x}{8} + \frac{Fl^3}{48}$$

Now substitute $y_{AB} = 0$ corresponding to $x = l$. The expression becomes

$$0 = -\frac{M_1 l^2}{2} + \frac{R_1 l^2}{6} - \frac{Fl^3}{6} + \frac{Fl^3}{4} - \frac{Fl^3}{8} + \frac{Fl^3}{48}$$

Simplifying and rearranging gives

$$M_1 - \frac{R_1 l}{3} = -\frac{Fl}{24}$$

7 Change the signs of the terms in Eq. (m) and add the result to Eq. (n). The M_1 terms cancel, and we have left

$$R_1 l - \frac{R_1 l}{3} = \frac{Fl}{2} - \frac{Fl}{24}$$

Solving for R_1 gives

$$R_1 = \frac{11F}{16}$$

Now substitute this result in Eq. (m) and solve for M_1. We then get

$$M_1 - \frac{11F}{16} l = -\frac{Fl}{2}$$

or $$M_1 = \frac{3Fl}{16}$$

8 Solving Eq. (a) for R_2 gives

$$R_2 = F - R_1 = F - \frac{11F}{16} = \frac{5F}{16}$$

Now that all the unknowns have been found, these may be substituted into Eqs. (*d*), (*e*), (*g*), and (*h*) to get the final expressions for the slope and deflection. When this is done, the results are found to be

$$\theta_{OA} = \frac{Fx}{32EI} (11x - 6l) \tag{q}$$

$$\theta_{AB} = \frac{F}{32EI} (10lx - 5x^2 - 4l^2) \tag{r}$$

$$y_{OA} = \frac{Fx^2}{96EI} (11x - 9l) \tag{s}$$

$$y_{AB} = \frac{F(l-x)}{96EI} (2l^2 - 10lx + 5x^2) \tag{t}$$

The maximum deflection will occur at a point along the beam where the slope is zero. It seems likely that this will occur to the right of *A*. Therefore, we set $\theta_{AB} = 0$ in Eq. (*r*) and solve for *x*. Canceling terms, we first obtain

$$10lx - 5x^2 - 4l^2 = 0$$

Changing signs, rearranging, and dividing by five gives

$$x^2 - 2lx + 0.8l^2 = 0 \tag{u}$$

From studies in algebra, we learn that the quadratic equation

$$ax^2 + bx + c = 0$$

can also be expressed as

$$x = \frac{-b \pm \sqrt{b^2 - 4ac}}{2a} \tag{v}$$

Solving Eq. (*u*) for *x* in this manner gives

$$x = \frac{2l \pm \sqrt{4l^2 - 3.2l^2}}{2} = l \pm 0.447l$$

Values of *x* greater than *l* have no meaning for us, and so we select the minus sign. Thus, the slope of the deflection curve is zero at

$$x = l - 0.447l = 0.553l \tag{w}$$

To get a formula for the maximum deflection, we simply place this value of *x* into Eq. (*t*) and simplify. This finally yields

$$y_{max} = -\frac{0.00932Fl^3}{EI} \tag{x}$$

8-5. BEAM FIXED AT BOTH ENDS   As a second example of the solution of indeterminate beam problems, we select one which occurs very frequently in design—a uniformly loaded beam with both ends fixed. In solving the problem,

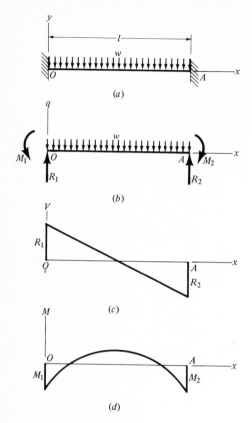

FIGURE 8-7.

we shall take advantage of symmetry wherever it is possible. The beam together with the loading, shear-force, and bending-moment diagrams are shown in Figs. 8-7*a* to *d*.

Because the beam and loading is symmetrical, the reactions R_1 and R_2 are equal. Therefore,

$$\Sigma F = R_1 - wl + R_2 = 0$$

And so

$$R_1 = R_2 = \frac{wl}{2} \tag{a}$$

The moment reactions M_1 and M_2 by symmetry are equal. As yet unknown, these must be obtained by requiring a zero slope at the fixed ends. So we next write Eq. (7-9) as

$$M = EI\frac{d^2y}{dx^2} = -M_1 + R_1x - wx\frac{x}{2} = -M_1 + \frac{wlx}{2} - \frac{wx^2}{2} \tag{b}$$

Integrating twice, as in the previous section, gives

$$EI\theta = EI\frac{dy}{dx} = -M_1x + \frac{wlx^2}{4} - \frac{wx^3}{6} + C_1 \qquad (c)$$

$$EIy = EIf(x) = -\frac{M_1x^2}{2} + \frac{wlx^3}{12} - \frac{wx^4}{24} + C_1x + C_2 \qquad (d)$$

The boundary conditions on Eq. (c) are that $\theta = 0$ at $x = 0$ and at $x = l$. From the first condition, we find that $C_1 = 0$. From the second condition, we get

$$-M_1l + \frac{wl^3}{4} - \frac{wl^3}{6} = 0$$

or $M_1 = \dfrac{wl^2}{12}$ $\qquad\qquad\qquad\qquad\qquad\qquad\qquad (e)$

The boundary conditions on Eq. (d) are that $y = 0$ at $x = 0$ and at $x = l$. From the first condition, we see that $C_2 = 0$.

If we now substitute C_1, C_2, and M_1 into Eq. (d) and simplify, we get for the deflection relation

$$y = -\frac{wx^2(l-x)^2}{24EI} \qquad (f)$$

By symmetry, y is maximum at the center of the beam. Using $x = l/2$ in Eq. (f) we get

$$y_{max} = -\frac{wl^4}{384EI} \qquad (g)$$

PROBLEMS

SECTION 8-1

8-1   A shouldered vertical steel shaft is supported at each end by thrust bearings, as shown schematically in the figure. Determine the proportion of the axial load W to be taken by each bearing.

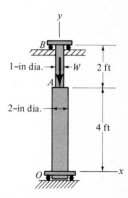

PROB. 8-1.

8-2 The figure shows a $\frac{3}{8}$ by $1\frac{1}{2}$ in rectangular steel bar welded to fixed supports at each end. The bar is axially loaded by the forces $F_A = 10\ 000$ lb and $F_B = 5000$ lb acting on pins at A and B. Assuming that the bar will not buckle laterally, find the reactions at the fixed supports.

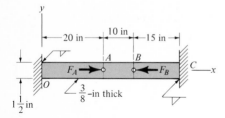

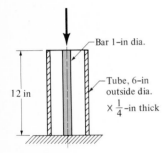

PROB. 8-2.

8-3 A compressive load whose resultant is $W = 30\ 000$ lb is supported by a steel bar 1 in in diameter and a 6-in diameter aluminum tube, as shown. Find the compressive stress in the tube and in the bar.

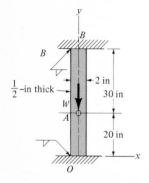

PROB. 8-3.

8-4 A rectangular aluminum bar $\frac{1}{2}$ in thick and 2 in wide is welded to fixed supports at the ends, and the bar supports a load $W = 800$ lb, acting through a pin, as shown. Find the reactions at the supports and the tensile and compressive stresses in the bar.

PROB. 8-4.

8-5 A bar 40 in long weighs 100 lb; it is to be supported by two springs, as shown. Where should spring k_2 be placed so that the bar will hang in a level position?

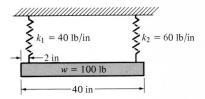

PROB. 8-5.

8-6 Two helical springs are connected in series, as shown. These are to be replaced by a single spring. What spring rate must it have in order that a force F would produce the same deflection?

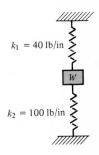

PROB. 8-6.

8-7 An electrical conductor is composed of a steel core $\frac{1}{16}$ in in diameter surrounded by a $\frac{3}{64}$ in shell of copper. What tension can be applied to the conductor if the stresses are not to exceed 60 kpsi and 24 kpsi for the steel and copper, respectively?

8-8 A weight $W = 80$ lb is suspended between two springs, as illustrated. Determine the forces carried by each spring.

PROB. 8-8.

8-9   Shown in the figure is a 20 lb rigid timber resting on two compression springs 12 ft apart. The timber supports a 180 lb weight on rollers. When the roller is located at some distance x from the left spring, the timber will be horizontal. Find x.

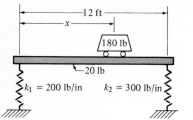

PROB. 8-9.

SECTION 8-2

8-10   The figure shows an assembly consisting of an aluminum tube and a steel bolt. The nut is first turned to the snug-tight position and then given an additional one-half turn. Find the stress in the tube and in the bolt.

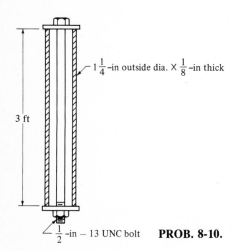

PROB. 8-10.

SECTION 8-3

8-11   The steel shaft shown in the figure is subjected to a torque T applied at point A. Find the torque reactions at O and B.

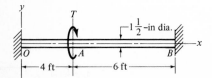

PROB. 8-11.

8-12   In testing the wear life of gear teeth, the gears are assembled using a pretorsion. In this way, a large torque can exist even though the power input to the tester is small. The arrangement shown in the figure uses this principle. Note the symbol used to indicate the location of the shaft bearings used in the figure. Gears A, B, and C are assembled first and then gear C is held fixed. Gear D is assembled and meshed with C by twisting it through an angle of 4° to provide the pretorsion. Find the maximum shear stresses in each shaft resulting from this preload.

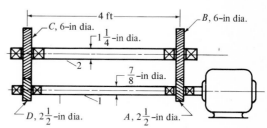

PROB. 8-12.

SECTIONS 8-4 AND 8-5

8-13 to 8-17   Verify the solutions to beams No. 12 through No. 15, respectively, in Table A-20.

SI UNITS

8-18   The figure shows a pressure cylinder that uses eight metric bolts having a proof strength of 620 MPa, as shown. These bolts are to be tightened to 80 percent of the proof strength.
(a) Find the tensile stress in the bolts and the compressive stress in the cylinder walls.
(b) Find the tensile stress in the bolts and the compressive stress in the cylinder walls after a fluid under a pressure of 60 kPa is introduced into the cylinder.

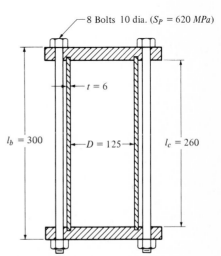

PROB. 8-18. All dimensions in millimetres.

columns

9-1. THE EULER PROBLEM   A short bar loaded in pure compression by a force P acting along the centroidal axis will shorten, in accordance with Hooke's law, until the stress reaches the elastic limit of the material. If P is increased still more, the material bulges and is squeezed into a flat disk.

Now visualize a long thin straight bar, such as a yardstick, loaded in pure compression by another force P acting along the centroidal axis. As P is increased from zero, the member shortens according to Hooke's law, as before. However, if the member is sufficiently long, as P increases, a critical value will be reached, designated as P_{cr}, corresponding to a condition of unstable equilibrium. At this point, any little crookedness of the member or slight movement of the load or support will cause the member to buckle and eventually collapse.

If the compression member is long enough to fail by buckling, it is called a *column*. A column failure can be a very dangerous failure because there is no warning that P_{cr} has been reached. In the case of a beam, an increase in the bending load causes an increase in the deflection; excessive deflection is always a visible indication of an overload. But, a column remains straight until the critical load is reached, after which there is sudden and total collapse. Depending upon the length, the actual stresses in a column at the instant of buckling may be quite low. For this reason, the material strengths, such as yield strength or ultimate strength, cannot be used to predict the degree of safety of a column.

Figure 9-1a shows a column of length l. The ends of the column are rounded or spherical, fitting into sockets in the supporting structure. Figure 9-1b is a free-body diagram of the column with the column load P replacing the effect of the supporting structure. Now assume that P is increased until the column just begins to buckle. If it happens to buckle in the positive y direction, then it will resemble Fig. 9-1c. Here, y is the distance from the x axis to the centroid of the cross section of the column. As shown, y depends upon x, and so

$$y = f(x) \qquad (a)$$

In fact, it is this particular function that we want to find.

The bar of Fig. 9-1c is bent in the positive y direction; this requires a negative bending moment, and so

$$M = -Py \qquad (b)$$

Note that a positive bending moment is needed to bend the bar in the negative y direction. But, since y is now negative, $M = -Py$, as before. If we now substitute Eq. (b) into Eq. (7-9), we obtain

$$\frac{d^2y}{dx^2} = -\frac{P}{EI}\, y \qquad (c)$$

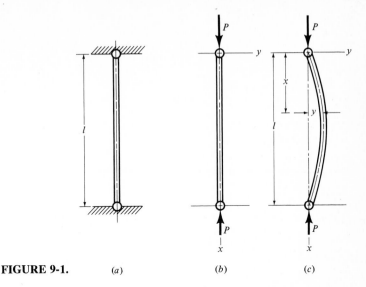

FIGURE 9-1.   (*a*)   (*b*)   (*c*)

The existence of the second derivative in this equation indicates that two in-tegrations are necessary to obtain y as a function of x. Unfortunately, y appears on the right side of the equation too, and so there is nothing to integrate. Equa-tion (*c*) is called a *differential equation;* such equations require special tech-niques for their solution. A simpler approach will be used here.

   If we examine Fig. 9-1*c* closely, the centroidal axis of the column seems to be bent into the form of a half trigonometric sine wave. This means that we think it may be possible that y (as a function of x) may have the form

$$y = A \sin \alpha x \tag{d}$$

where both A and α are constants yet to be determined. To test this supposition, we use Table A-5 twice to find the first and second derivatives of Eq. (*d*). The first derivative is

$$\frac{dy}{dx} = \frac{d}{dx} (A \sin \alpha x) = A \cos \alpha x \frac{d(\alpha x)}{dx} = A\alpha \cos \alpha x \tag{e}$$

The second derivative is

$$\frac{d^2y}{dx^2} = \frac{d}{dx} (A\alpha \cos \alpha x) = -A\alpha \sin \alpha x \frac{d(\alpha x)}{dx} = -A\alpha^2 \sin \alpha x \tag{f}$$

Now substitute Eqs. (*d*) and (*f*) into Eq. (*c*). This gives

$$-A\alpha^2 \sin \alpha x = -\frac{P}{EI} A \sin \alpha x \tag{g}$$

Here we can cancel $A \sin \alpha x$ from both sides of the equation, leaving

$$\alpha^2 = \frac{P}{EI}$$

Thus, Eq. (*d*) is indeed a solution. Replacing α, Eq. (*d*) becomes

$$y = A \sin \sqrt{\frac{P}{EI}} \, x \qquad\qquad (h)$$

To evaluate A, we simply observe that the boundary conditions require that $y = 0$ when $x = l$. This means that[1]

$$A \sin \sqrt{\frac{P}{EI}} \, l = 0 \qquad\qquad (i)$$

The trivial solution of no buckling occurs with $A = 0$. However, if A is not zero, then Eq. (*i*) requires that

$$\sin \sqrt{\frac{P}{EI}} \, l = 0 \qquad\qquad (j)$$

Angles in degrees for which the sine function is zero are 0, 180, 360, etc. In radian measure, these angles may be designated as $N\pi$, where $N = 0, 1, 2$, etc. Using $N = 1$ gives

$$\sqrt{\frac{P_{cr}}{EI}} \, l = \pi$$

or $\qquad P_{cr} = \dfrac{\pi^2 EI}{l^2} \qquad\qquad (9\text{-}1)$

This is the famous *Euler column formula*. Substituting P_{cr} for P in Eq. (*h*) gives

$$y = A \sin \sqrt{\frac{\pi^2 EI}{l^2} \frac{1}{EI}} \, x = A \sin \frac{\pi x}{l} \qquad\qquad (k)$$

If we should calculate a set of values of y from this equation for values of x between $x = 0$ and $x = l$, we would get a half sine wave like Fig. 9-1c upon plotting the results. We would also find that $y_{max} = A$ at $x = l/2$.

The critical load occurs with $N = 1$. Higher values of N result in deflection curves that cross the x axis at points of inflection and are multiples of half sine waves. They also require higher values of P.

9-2. FORMULAS FOR LONG COLUMNS   Euler's formula can be placed in a more convenient form by introducing *radius of gyration* k into the equation. Using $I = Ak^2$ (see Appendix A-8) in Eq. (9-1) and rearranging gives

$$\frac{P_{cr}}{A} = \frac{\pi^2 E}{(l/k)^2} \qquad\qquad (9\text{-}2)$$

---

[1] It can also be shown that another solution to Eq. (*c*) is

$$y = B \cos \sqrt{\frac{P}{EI}} \, x$$

However, this solution is of no value since, with $y = 0$ at $x = 0$, we find that the coefficient $B = 0$.

Here, the term l/k is called the slenderness ratio. And the term P_{cr}/A is called the *critical unit load*. Though the unit load has the same dimensions as stress, you should *never* call it a stress! To do so might lead to the error of comparing it with a strength, the yield strength, say, and to the false conclusion that a margin of safety exists.

Equation (9-2) shows that the critical unit load depends *only* upon the modulus of elasticity and the slenderness ratio. Thus, an Euler column need never be made from a high-strength alloy steel because a lower-priced and lower-strength steel will have the same modulus of elasticity.

Equation (9-2) applies to any column with both ends rounded or pinned. This is the equation that is used the most by designers for long columns because it is usually safer to assume that the ends are pinned. Figure 9-2 shows some of the other end conditions that may be encountered. Equations for the critical loads for these columns may be obtained by solving Eq. (*c*) of Sec. 9-1, using the appropriate boundary conditions. Another, and simpler, approach is to compare the deflection curves of Fig. 9-2 with the half-sine-wave configuration of Fig. 9-1.

The column of Fig. 9-2a, with both ends fixed, has inflection points at A and B.[1] The portion of the column between these points has a half-sine-wave configuration and is of length $l/2$. Substituting this length into Eq. (9-1) gives

$$P_{cr} = \frac{\pi^2 EI}{(l/2)^2} = \frac{4\pi^2 EI}{l^2} \qquad (9\text{-}3)$$

as the critical load for a column with both ends fixed.

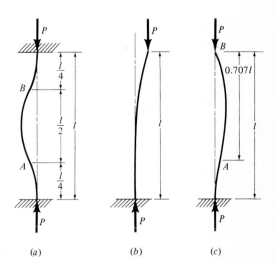

(a)   (b)   (c)

FIGURE 9-2. Column end conditions. (*a*) Both ends fixed; (*b*) one end fixed, one end free; (*c*) one end fixed, one end rounded and guided.

[1] An inflection point is a point on a curve where the curve changes from a concave to a convex configuration.

TABLE 9-1
Theoretical end condition constants *n*
for Euler columns.

end conditions	n
Fixed-free	$\frac{1}{4}$
Rounded-rounded	1
Fixed-guided	2
Fixed-fixed	4

Figure 9-2*b* shows a column with one end fixed and one end free. The deflection curve of the bent column is the same as the curve for one-half of a column with both ends rounded. This means that the length *l* of the column of Fig. 9-2*b* corresponds to *l*/2 for the column with rounded ends. Therefore, we replace *l* in Eq. (9-1) with 2*l* to get

$$P_{cr} = \frac{\pi^2 EI}{(2l)^2} = \frac{\pi^2 EI}{4l^2} \tag{9-4}$$

A column with one end fixed and the other end rounded and guided is shown in Fig. 9-2*c*. An inflection point occurs at *A*, a distance of 0.707*l* from the rounded end. The portion of the column from *A* to *B* has the same bending configuration as the entire length of a column with both ends rounded. Thus, we replace *l* in Eq. (9-1) by 0.707*l* to get

$$P_{cr} = \frac{\pi^2 EI}{(0.707l)^2} = \frac{2\pi^2 EI}{l^2} \tag{9-5}$$

It is convenient to write both forms of the Euler equations as

$$P_{cr} = \frac{n\pi^2 EI}{l^2} \qquad \frac{P_{cr}}{A} = \frac{n\pi^2 E}{(l/k)^2} \tag{9-6}$$

where *n* is called the *end-condition constant*. For convenience, these values found above are listed in Table 9-1. These are called *theoretical* constants because many designers feel that it is too dangerous to use the higher values of *n*, even though they are theoretically correct.

Columns that obey the Euler formula are called either *Euler columns* or *long columns*. The reason for this is not hard to deduce. As we make any compression member shorter and shorter, we eventually get a member so short that it will not fail by buckling. That is, it will fail by mashing or squashing instead. Thus, there must exist some length below which a compression member is not a column. In the next section, we explore this question in greater detail.

9-3. COLUMN FAILURE   Two possible criteria for the failure of a compression member are that the stress may exceed the strength of the material, or that the member may fail by buckling. It is easy to diagram both of these methods on a graph, and this has been done in Fig. 9-3*a*. Using the unit load *P/A* as the ordinate and the slenderness ratio as the abscissa enables us to plot a graph of the

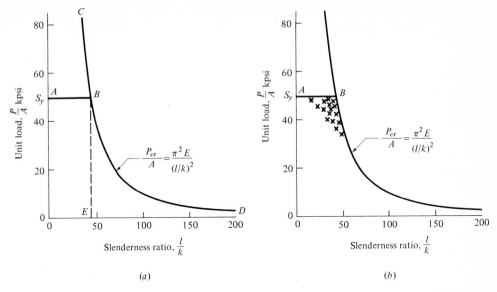

FIGURE 9-3. (*a*) **Possible failure criteria for compression members;** (*b*) **same graph, but *x*'s mark typical experimental column failures.**

Euler equation for a column with both ends rounded. This curve is identified as *CBD* in the figure. Also, since yield strength is often the criteria used for a simple compression failure, the horizontal line *AB* represents the upper limit of the unit load *P/A*. Thus, points on this plot represent failure if they fall outside the graph *ABD*, or safety if they fall inside. The advantage of this graph is that it seems to distinguish between a column and a pure compression member. Thus, any member having a slenderness ratio to the left of *E* in Fig. 9-3*a* is a pure compression member; such a member would be analyzed for safety by comparing the compressive stress with the strength. But if the slenderness ratio fell to the right of point *E*, the member would be classified as a column; then safety would be indicated if the column load *P* were less than the buckling load P_{cr}. Unfortunately, this beautiful theory doesn't hold up in practice!

Figure 9-3*b* is the same graph, in which the *x*'s added to the graph mark typical experimental column failure points. These are seen to be clustered around point *B*, where one failure criteria intersects the other. People who have studied this problem in detail say that the reason the theory is not verified by experiment near *B* is that it is practically impossible to construct an ideal column. Almost any member will have slight deviations from straightness, and it is just as difficult to load the ends perfectly so as to achieve the idealized end conditions. Any very small deviations can have an enormous effect upon the buckling load. So such factors as residual stresses, initial crookedness, and slight load eccentricities all contribute to the scatter of the failure points near *B*, and hence the deviation from the theory.

Of course, all of this leaves us just as much up in the air as before, since we still are not able to distinguish a column from a pure compression member. So, in the next several sections, we shall learn how engineers have coped with this problem.

9-4. THE PARABOLIC FORMULA   Many different column formulas, most of them empirical, have been devised to overcome some of the disadvantages of the Euler equation. The *parabolic, or J. B. Johnson, formula* is widely used in the machine, automotive, aircraft, and structural-steel construction fields. This formula often appears in the form

$$\frac{P_{cr}}{A} = a - b \left(\frac{l}{k}\right)^2 \tag{9-7}$$

where a and b are constants that are adjusted to cause the formula to fit experimental data. Figure 9-4 is a graph of two formulas, the Euler equation and the parabolic equation. Notice that curve ABD is the graph of the parabolic formula, while curve BC represents the Euler equation. In analyzing a column to determine the critical buckling load, only part AB of the parabolic graph and part BC of the Euler graph should be used.

The constants a and b in Eq. (9-7) are evaluated by deciding where the intercept A, in Fig. 9-4, is to be located, and where the tangent point B is desired. Notice that the coordinates of point B are specified as $(P/A)_1$ and $(l/k)_1$.

One of the most widely used versions of the parabolic formula is obtained by making the intercept A correspond to the yield strength S_y of the material, and making the parabola tangent to the Euler curve at $(P/A)_1 = S_y/2$. Thus, the

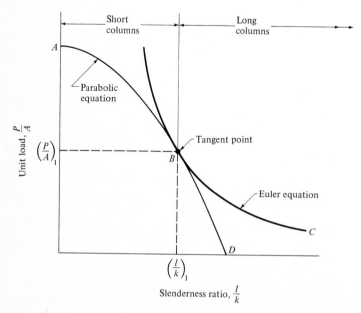

FIGURE 9-4.

first constant in Eq. (9-7) is $a = S_y$. To get the second constant, substitute $S_y/2$ for P_{cr}/A and solve for $(l/k)_1$. Using Eq. (9-6) we get

$$\left(\frac{l}{k}\right)_1 = \sqrt{\frac{2\pi^2 nE}{S_y}} \tag{9-8}$$

Then, substituting all this into Eq. (9-7) yields

$$\frac{S_y}{2} = S_y - b\,\frac{2\pi^2 nE}{S_y}$$

or $$b = \left(\frac{S_y}{2\pi}\right)^2 \frac{1}{nE} \tag{9-9}$$

which is to be used for the constant in

$$\frac{P_{cr}}{A} = S_y - b\left(\frac{l}{k}\right)^2 \tag{9-10}$$

Of course, this equation should only be used for slenderness ratios up to $(l/k)_1$. Then the Euler equation is used when l/k is greater than $(l/k)_1$.

Equations (9-6) and (9-9) have been solved using the end conditions of Table 9-1 for AISI 1015 cold-drawn steel, and the results are plotted in Fig. 9-5. This is called a design chart. The examples that follow show how to obtain such charts for any material and end conditions, and how to use them in analysis and design.

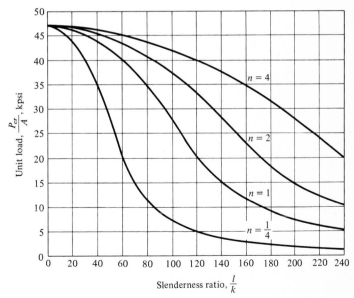

FIGURE 9-5. Design chart using parabolic equation and Euler equation for AISI 1015 cold-drawn steel for theoretical end conditions.

Sometimes, the parabolic equation is derived with the factor of safety included. For example, the AISC column formula for ASTM No. A7 steel (see Table A-19) for both main and secondary members is

$$\frac{P}{A} = 17\,000 - 0.485 \left(\frac{l}{k}\right)^2 \tag{9-11}$$

which is valid only for l/k values less than 120. Since A7 steel has a tensile strength of 60 kpsi or more, this equation has a factor of safety built in. Note that this fact is also evident because it has been written in terms of P/A, instead of P_{cr}/A.

EXAMPLE 9-1   Derive the parabolic column formula for AISI 3140 hot-rolled steel for a column with rounded ends. Find the tangent point with the Euler equation and plot a graph of P_{cr}/A vs. the slenderness ratio for l/k values up to 200.

SOLUTION

From Table A-18 we find $S_y = 64$ kpsi. Also, $n = 1$ and $E = 30$ Mpsi for steel. The constant b is found, from Eq. (9-9), to be

$$b = \left(\frac{S_y}{2\pi}\right)^2 \frac{1}{nE} = \left[\frac{64(10)^3}{2\pi}\right]^2 \frac{1}{1(30)(10)^6} = 3.46$$

Then, Eq. (9-10) becomes

$$\frac{P_{cr}}{A} = 64(10)^3 - 3.46 \left(\frac{l}{k}\right)^2 \qquad Ans.$$

Next, from Eq. (9-8), we get the tangent point as

$$\left(\frac{l}{k}\right)_1 = \sqrt{\frac{2\pi^2 nE}{S_y}} = \sqrt{\frac{2\pi^2(1)(30)(10)^6}{64(10)^3}} = 96.2 \qquad Ans.$$

Thus, we use the parabolic equation for l/k's up to 96.2 and the Euler equation beyond.

For example, using $l/k = 60$, we get the unit buckling load as

$$\frac{P_{cr}}{A} = 64(10)^3 - 3.46(60)^2 = 51\,500 \text{ psi}$$

But, if l/k is 120, the unit buckling load is

$$\frac{P_{cr}}{A} = \frac{n\pi^2 E}{(l/k)^2} = \frac{(1)(\pi)^2(30)(10)^6}{(120)^2} = 20\,600 \text{ psi}$$

By computing other values of P_{cr}/A in the same way, we obtain the graph of Fig. 9-6.   ////

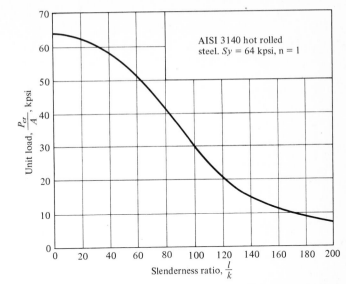

FIGURE 9-6. Design chart for AISI 3140 hot-rolled steel columns.

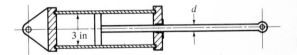

FIGURE 9-7.

EXAMPLE 9-2   The hydraulic cylinder shown in Fig. 9-7 has a 3 in bore and is to operate at a pressure of 800 psi. With the clevis mount shown, the piston rod should be sized as a column with both ends rounded. The rod is to be made of 3140 hot-rolled steel, using a factor of safety of 3. So use the design chart of Fig. 9-6.

(a) Find the piston rod diameter d if the column length is 60 in.

(b) Find d if $l = 18$ in.

SOLUTION

(a) The column load is

$$P = \frac{\pi D^2}{4} p = \frac{\pi (3)^2}{4} (800) = 5650 \text{ lb}$$

Therefore, for a factor of safety of 3, the column is to be designed for a critical load of

$$P_{cr} = 3P = 3(5650) = 17\,000 \text{ lb}$$

We shall solve this problem by guessing at a diameter. Corresponding to this guess, we calculate the slenderness ratio and the unit load. We then enter the design chart of Fig. 9-6 with these two parameters. If they locate a point below the design curve, the design is safe. But if the point is above, the design is not safe. Of course, the object is to find a diameter which will define a point as close to the design curve as possible, without going over.

First guess: $d = 1\frac{1}{2}$ in. Then, from Table A-15,

$$k = \frac{d}{4} = \frac{1.5}{4} = 0.375 \text{ in}$$

The slenderness is now found as $l/k = 60/0.375 = 160$. The cross-sectional area is $A = \pi d^2/4 = \pi (1.5)^2/4 = 1.77 \text{ in}^2$. And so the unit load is

$$\frac{P_{cr}}{A} = \frac{17\ 000}{1.77} = 9600 \text{ psi}$$

We now enter Fig. 9-6 with $P_{cr}/A = 9600$ psi and $l/k = 160$. This yields a point below the design curve, and so a $1\frac{1}{2}$ in diameter column is safe.

Second guess: $d = 1\frac{3}{8}$ in. Then $k = d/4 = 1.375/4 = 0.344$. Also, $l/k = 60/0.344 = 174$. And $A = \pi (1.375)^2/4 = 1.48 \text{ in}^2$. So

$$\frac{P_{cr}}{A} = \frac{17\ 000}{1.48} = 11\ 500 \text{ psi}$$

If we now enter the chart again with $P_{cr}/A = 11\ 500$ psi and $l/k = 174$, we locate a point above the design line. So this design is not safe. Thus, if we wish a diameter to the nearest $\frac{1}{8}$ in, the correct result is $1\frac{1}{2}$ in.

Note that we used the Euler portion of the design curve to get this result. Had we known this in advance, we could have solved the first part of Eq. (9-6) directly for the column diameter. Try it.

(b) Using the same procedure as in (a), we first guess that $d = \frac{3}{4}$ in. This gives $k = 0.1875$ in, $l/k = 96$, $A = 0.441 \text{ in}^2$, and $P_{cr}/A = 38.5$ kpsi. When we enter the chart with these figures, we locate a point above the design line, and so a $\frac{3}{4}$ in diameter is not safe. But, using $d = \frac{7}{8}$ in, we get $l/k = 82$ and $P_{cr}/A = 28.3$ kpsi, which is well below and hence is safe. In this case, the parabolic portion of the design curve was used, and so the solution must be obtained by trial and error.   ////

EXAMPLE 9-3 Determine the safe compressive load that can be carried by a hot rolled structural steel angle 6 by 6 by $\frac{3}{4}$ in and 8 ft long.

SOLUTION

From Table A-10, we find that $A = 8.44 \text{ in}^2$ and that the minimum radius of gyration is $k = 1.17$. Therefore, $l/k = (8)(12)/1.17 = 82$. Since this is less than 120, Eq. (9-11) applies. Therefore,

$$\frac{P}{A} = 17\ 000 - 0.485(82)^2 = 13\ 700 \text{ psi}$$

And so the safe load P is

$$P = (13\ 700)A = (13\ 700)(8.44) = 115\ 600\ \text{lb} \quad Ans.$$ ////

9-5. THE GORDON-RANKINE FORMULA

An empirical formula, first proposed by Gordon, and later revised by a colleague, Rankine, is expressed in general form as

$$\frac{P}{A} = \frac{\sigma_{max}}{1 + \phi(l/k)^2} \tag{9-12}$$

where σ_{max} is the maximum compressive stress to be permitted in the column and ϕ is a constant that is adjusted to cause the formula to fit experimental data. For very short compression members, the slenderness ratio is quite small and Eq. (9-12) becomes, approximately,

$$\frac{P}{A} \cong \sigma_{max}$$

But for long columns, l/k is large and then the equation approximates

$$\frac{P}{A} \cong \frac{\sigma_{max}}{\phi(l/k)^2}$$

which is the same form as the Euler equation. Thus, the Gordon-Rankine formula represents an attempt to obtain a single formula useful for all slenderness ratios. In practice, it is difficult to select values of σ_{max} and ϕ to produce satisfactory results for all column lengths. Consequently, the range of l/k values is usually specified after the constants have been specified.

A particular example of the use of Eq. (9-12) is the AISC formula for bracing and secondary structural-steel members. The constants are $\sigma_{max} = 18\ 000$ and $\phi = 1/18\ 000$, and so the equation is

$$\frac{P}{A} = \frac{18\ 000}{1 + [(l/k)^2/18\ 000]} \tag{9-13}$$

for l/k values greater than 120. Remember, that for l/k less than 120, Eq. (9-11) applies.

For main members, Eq. (9-13) is modified by the AISC code to

$$\frac{P}{A} = \frac{18\ 000}{1 + [(l/k)^2/18\ 000]}\left(1.6 - \frac{l/k}{200}\right) \tag{9-14}$$

These equations have the factor of safety built in and hence give the safe loads. The material is ASTM A7 steel.

9-6. OTHER COLUMN FORMULAS

Formulas like

$$\frac{P}{A} = \sigma_o - C\frac{l}{k} \tag{9-15}$$

where σ_o is either a strength or a safe stress and C is a constant that depends upon the column ends, have been used in the past. Such formulas give results that are only a rough approximation to experimental results. We shall not use them in this book; they have been mentioned only for their historical interest. The solution speed made available by the compact pocket calculator means that better results can be obtained faster with the more accurate approaches, such as the parabolic formula.

Most compression members are not perfectly straight, and it is also rather difficult to assure that column loads are exactly concentric. Thus, one might as well admit that a certain amount of eccentricity and crookedness is always present and introduce these effects into the formulas. When this is done, the secant formula is the result. It is a rational formula, and it is also valid for all column lengths. It turns out, however, that it is quite difficult to use in design, though the use of calculators does shorten the work. Another characteristic is that the secant formula gives the stress, instead of the unit buckling load. There may not, in fact, be any simple relation between stress and buckling.

The secant formula may be derived by using appropriate boundary conditions in the general solution to Eq. (c) of Sec. 9-1. The result, after some manipulation, is

$$\sigma_{\max} = \frac{P}{A}\left(1 + \frac{ec}{k^2}\ sec\ \frac{l}{k}\ \sqrt{\frac{P}{4AE}}\ \right) \tag{9-16}$$

which is the usual form of the *secant formula*. Here, e is the initial eccentricity or crookedness of the column. Thus, e is the distance between the centroidal axis of the column and the axis along which the load is applied. The distance from the neutral axis of the section to the surface under greatest compressive bending stress is denoted by the distance c. The remaining symbols have been previously identified.

It is important to note that the relationship between the stress and the load in Eq. (9-16) is not linear. Thus, for design purposes, apply the factor of safety to the column load, not to the stress. Equation (9-16) can be used for design purposes by substituting S_y for $\sigma_{\max}$ and $P_{\max}$ for P. Then, $P_{\max}$ is the load that would produce a column failure.

Local buckling is another interesting example of instability, though not necessarily a column problem. If we were to place an empty beer can between two short boards or plates and squeeze them between the jaws of a wood clamp, we would have produced a simple experiment for demonstrating local buckling. Tightening up the clamp will cause a failure or wrinkling of the sides of the can. Depending upon the concentricity of the load, we will usually get some local buckling in which part of the can bulges in and part of it bulges out. Other kinds of local failure may involve torsion and bending too. While these make interesting mathematical problems, we shall not consider them here.

PROBLEMS

9-1 Suppose the sine function of Eq. (j) of Sec. 9-2 is 2π. Determine P_{cr} and plot the configuration of a buckled column using $l = 5$ in and $A = 0.5$ in.

9-2 Steel bars 3 ft long with rounded ends are to be treated as columns. Find the slenderness ratio and the buckling load if:

(a) A bar has a rectangular cross section $\frac{1}{4}$ by 1 in.

(b) A bar has a square cross section $\frac{1}{2}$ by $\frac{1}{2}$ in.

(c) A bar is round and has a diameter of 0.564 in.

9-3 A rectangular bar, to be used as a column, is to be fixed in one plane of buckling and rounded in the other. In order to get the strongest column, and based on theoretical end conditions, what should be the ratio of the width to the thickness of the cross section?

9-4 Structural steel shapes 10 ft in length are to be treated as columns using an end-condition constant $n = 1$. The shapes which follow have approximately the same weight per foot. Find the critical buckling load for each.

(a) An angle $2\frac{1}{2}$ by $2\frac{1}{2}$ by $\frac{3}{8}$ in.

(b) An angle 4 by 3 by $\frac{1}{4}$ in.

(c) A 3-in 6-lb channel.

(d) A 3-in 5.7-lb I beam.

(e) A steel tube $2\frac{1}{4}$ in OD with a wall thickness of $\frac{9}{32}$ in.

9-5 A Euler column with one end fixed and one end free is to be made of an aluminum alloy. It is to have a cross-sectional area of 1 in² and a length of 96 in. Determine the column buckling load corresponding to the following shapes:

(a) A solid round bar.

(b) A hollow round bar with a 2 in OD.

(c) A square bar.

(d) A hollow square bar with a 2 in outside dimension.

9-6 Derive parabolic formulas such that the intercept with the P/A axis is the yield strength, and the tangent with the Euler curve is half the yield strength, for the following cases:

(a) Aluminum alloy No. 2024-T3; one end fixed, one end guided.

(b) AISI 1010 hot-rolled steel; one end fixed, one end free.

(c) AISI 3140 steel heat treated and drawn to 800 F; both ends rounded.

9-7 Derive the parabolic formula for a column, with both ends rounded, and made of AISI 1018 cold-drawn steel. Place the intercept with the unit load at the yield strength, and the tangent point to the Euler curve at half the yield strength. Make a combined graph of the unit load vs. the slenderness ratio out to $l/k = 250$. Save this graph for use in some of the problems which follow.

9-8   A $1\frac{1}{2}$ in standard steel pipe has an OD of 1.900 in and an ID of 1.610 in. Find the critical load for a 3-ft length of this pipe, assuming it is made of 1018 cold-drawn steel and using the chart of Prob. 9-7.

9-9   A solid round steel bar is to be designed to support a compressive load of 1200 lb. The member is to be designed with rounded ends using cold-drawn 1018 steel. Use a factor of safety of 4 and the design chart of Prob. 9-7. Determine the safe diameter to the nearest $\frac{1}{8}$ in for column lengths of 4 in, 12 in, and 36 in.

9-10   A steel tube having a $\frac{3}{16}$-in wall thickness is to be designed to support a column load of 2800 lb. The column will have the ends pivoted and will be made of cold-drawn 1018 steel. Use the design chart of Prob. 9-7, a factor of safety of 3.5, and find the outside diameter to the nearest $\frac{1}{8}$ in for the following lengths:
  (a) $l = 3$ in
  (b) $l = 15$ in
  (c) $l = 45$ in

9-11   The 500-lb load at E, on the linkage illustrated, places link AD in compression. This link is 2 in wide and $\frac{3}{8}$ in thick. Due to the compressive load, the link could buckle in two ways—in the plane of the linkage or out of the plane. The theoretical end conditions are fixed-fixed for one of the modes, and rounded-rounded for the other. The link is made of cold-drawn 1015 steel. Use Fig. 9-5 and find the factor of safety guarding against buckling. Assume theoretical end conditions apply. Use Euler's equation in case the slenderness ratio is greater than 240.

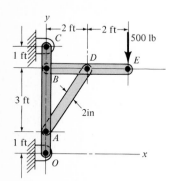

PROB. 9-11.

9-12   The figure shows a 12 ft cantilevered linkage rigged to support a weight $W = 4000$ lb. Since link 3 is subject to column action, it is to be designed using a factor of safety of 5 and AISI 1015 cold-drawn steel. Use cross-sectional proportions $w = 6t$. Note that the ends are pivoted (rounded) for buckling parallel to the xy plane and fixed for buckling in the z direction. Use Fig. 9-5, if applicable, and find safe cross-sectional dimensions to the nearest $\frac{1}{8}$ in.

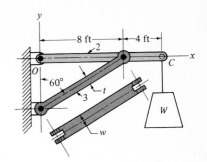

PROB. 9-12.

9-13 The figure shows a double-angle strut often used for bracing in steel construction. If the angles are 6 by 6 by $\frac{3}{4}$ in, find the safe column loads for:

(a) An unsupported length of 16 ft

(b) An unsupported length of 30 ft

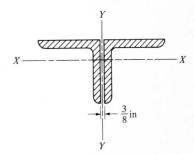

PROB. 9-13.

9-14 The same as Prob. 9-13 except the angles are 4 by 4 by $\frac{1}{2}$ in for:

(a) An unsupported length of 8 ft

(b) An unsupported length of 30 ft

9-15 The figure shows the cross section of a double-angle strut which is used for bracing and secondary compression members in steel construction. Find the safe column loads for both directions of buckling if the angles are 8 by 4 by $\frac{1}{2}$ in and the unsupported length is 16 ft.

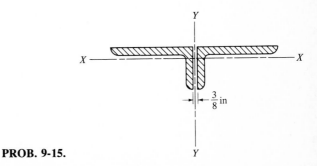

PROB. 9-15.

9-16   The same as Prob. 9-15, except the angles are 6 by 4 by $\frac{3}{4}$ in and the unsupported length is 18 ft.

9-17   A 14 by 16 WF 228 shape is to be used as a main column with an unsupported length of 30 ft. Compute the allowable concentric load.

SI UNITS

9-18   A solid round steel bar is to be designed to support a compressive load of 6000 N. The member is to be designed with pivoted ends using a steel having a yield strength of 360 MPa. Use a factor of safety of 3.8 and determine the safe diameter to the nearest millimetre for column lengths of 0.1 m, 0.3 m, and 1 m.

combined stresses

10-1. INTRODUCTION In this chapter, we are going to study what happens when two or more stresses are present at the same point in a member. One of the most common examples of this is a geared rotating shaft. The gears transmit torsion to the shaft and they also cause the shaft to bend. Thus, at the same point, we can expect to have both a shear stress (due to the torsion) and a normal stress (due to the bending). The stresses resulting from any such combination of loads are called *combined stresses*.

There are only five basic stresses, three shear stresses and two normal stresses, with which we shall be concerned here. Two axial stresses were developed in Chap. 2. From Eq. (2-1), the normal axial stress in tension or compression is

$$\sigma = \pm \frac{F}{A} \tag{10-1}$$

And, from Eq. (2-2), the direct shear stress is

$$\tau = \frac{F}{A} \tag{10-2}$$

Two basic bending stresses have been investigated. From Eq. (5-3), the normal bending stress is

$$\sigma = \frac{My}{I} \tag{10-3}$$

The shear stress due to bending, from Eq. (5-8), is

$$\tau = \frac{VQ}{Ib} \tag{10-4}$$

Finally, the torsional shear stress, expressed as Eq. (6-2), is

$$\tau = \frac{T\rho}{J} \tag{10-5}$$

So it is the combinations of these five stresses with which we shall be concerned in this chapter.

It is necessary to use the *principle of superposition* in combining stresses. This principle applies only when the stresses are linearly related to the forces which produced them. Thus, columns are excepted because the stress in a column is not directly proportional to the load.

The principle of superposition for stresses states that *stresses of the same kind may be summed algebraically to obtain the resultant stress.* Thus, two

normal stresses acting in the same direction may be added; if they act in opposite directions, they are subtracted. Similarly, shear stresses acting in the same
direction are added; they are subtracted to get the resultant if they act in opposite directions. The advantage of the principle of superposition is that the
same kind of stresses (shear or normal) may be summed regardless of how the
stresses were produced.

10-2. NORMAL STRESS COMBINATIONS   The connecting rod of Fig. 10-1a is
subjected to an axial tension due to the force F, which is applied through pins at
the ends, and a bending due to its own weight. Figure 10-1b is a free-body
diagram of the rod showing the uniform bending load w and the tensile forces F.
If the cross section is rectangular, as in Fig. 10-2a, then the distances c from the
neutral axis to each outer surface are equal. Due to the tensile load, we get a tensile stress σ_a (Fig. 10-2b). Due to the bending, we get a stress as given by Eq.
(10-3). Superposing these two stresses gives

$$\sigma = \frac{F}{A} \pm \frac{My}{I} \qquad\qquad (a)$$

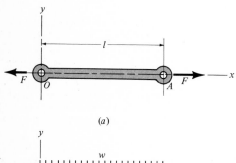

(a)

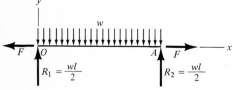

(b)   **FIGURE 10-1.**

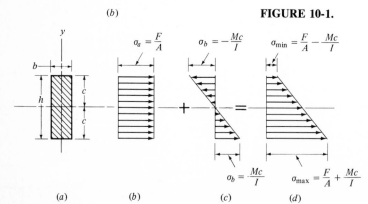

(a)   (b)   (c)   (d)   **FIGURE 10-2.**

The stress distribution corresponding to this equation is shown in Fig. 10-2d. Since the bending produces tension in the bottom surface, the maximum stress occurs there, as shown in Fig. 10-2d. Also shown is the minimum stress, still tension in this example, at the top surface.

Another example is furnished by the short compression member in Fig. 10-3a. Here a T section is loaded in compression by a force F not on the centroidal axis of the section. This is an example of *eccentric loading*. The distance e from the centroid of the section (Fig. 10-3b) to B, the point of application of the load, is called the *eccentricity*. Figure 10-3c is a free-body diagram of the member. The force reaction $R_1 = F$ acts at the centroid O. Since R_1 and F form a force couple having an arm equal to the eccentricity, a moment reaction $M_1 = Fe$ must also act at the support.

Figure 10-4 shows the stress distributions composed of the axial stress, the bending stress, and the combination or resultant of the two. The axial stress σ_a is a uniform compressive stress, as shown in Fig. 10-4b. The bending stress σ_b has the familiar triangular distribution, shown in Fig. 10-4c, with the bottom of the T

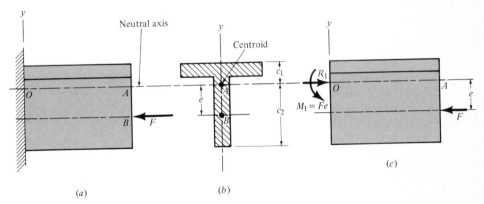

(a) (b) (c)

FIGURE 10-3.

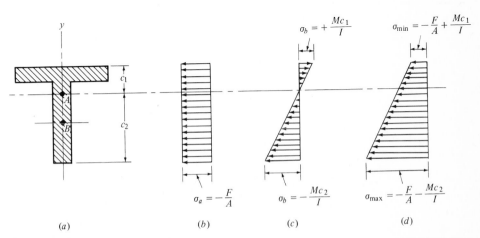

(a) (b) (c) (d)

FIGURE 10-4.

in compression and the top surface in tension. These two distributions are superposed in Fig. 10-4d. The resultant stress distribution is given by the equation

$$\sigma = -\frac{F}{A} \pm \frac{My}{I} \qquad (b)$$

where y must be measured from the neutral axis. The maximum compressive stress occurs at the bottom and the minimum occurs on top, as shown.

10-3. SHEAR STRESS COMBINATIONS   The principle of superposition applies to the problem of combining shear stresses too. Equation (6-15) is an application of the principle to the determination of the maximum shear stress in a helical spring, which we have already studied. An extension of the same principle was used in Sec. 6-7 in studying the torsion in welded joints. Here, primary and secondary shear stresses not having the same direction were combined by the use of the pythagorean theorem.

10-4. THE USE OF STRESS ELEMENTS   Stress elements were used in Sec. 6-4 to study the principal stresses in torsion. We now wish to study a case of torsion combined with bending, which is much more general. In Fig. 10-5a is shown the free-body diagram of a countershaft loaded by a force F_1 and a torque T_1 at O, and by another force F_2 and torque T_2 at D. We assume, too, that bearings placed at A and C support the shaft through bearing reactions R_1 and R_2. No torque reaction is needed because $T_1 = T_2$.

A stress element on the side of the surface at B in Fig. 10-5a will have a tensile bending stress of magnitude

$$\sigma_x = \frac{Mc}{I} \qquad (a)$$

and a torsional shear stress

$$\tau_{xy} = \frac{Tr}{J} \qquad (b)$$

Now, in Fig. 10-5b, this stress element has been removed and shown in a three-dimensional form properly aligned with respect to the xyz axes. The faces of the stress element are named according to the axes which are normal to them. Thus, the element has a positive x face, which is the face normal to the x axis. It also, has a negative x face, which is opposite to the positive x face.

The stress τ_{xy} is parallel to the x face and in the y direction. The stress τ_{yx} is parallel to the y face and in the x direction. Static equilibrium requires that

$$\tau_{yx} = \tau_{xy} \qquad (d)$$

The normal stresses are subscripted according to the axes to which they are parallel. Thus, a general stress element could have a normal stress σ_y, parallel to y, and even a normal stress σ_z, parallel to z. Note that the stress element

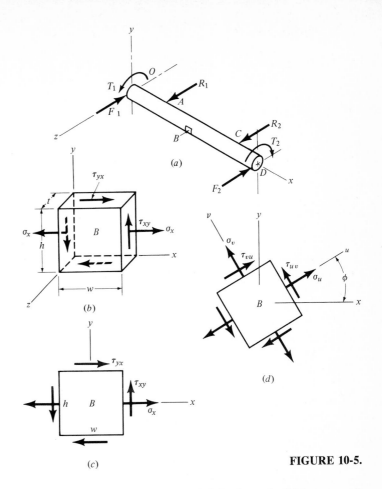

FIGURE 10-5.

has also been given the dimensions of thickness t, height h, and width w to call attention to the fact that the size of the stress element is of importance.

In most cases of interest to us, there is no stress in the z direction, and so it isn't necessary that we diagram stress elements in all three of their dimensions. Thus, Fig. 10-5c is the same stress element. It is called a *plane stress element* because we can think of it as existing in a single plane. Then, the stresses shown acting on the element are called *plane stresses*. We must not forget, however, that the element still has a thickness t, even though it doesn't show on the diagram.

We now ask the question: What would happen if we choose a plane stress element having sides not parallel to the xy axes? Suppose, in fact, that we are interested in the element of Fig. 10-5d. This element has a set of uv axes at an angle ϕ to the xy system. Since it is a plane stress element, we might reasonably expect it to have normal stresses σ_u and σ_v, as well as shear stresses τ_{uv} and τ_{vu}. Since the element of Fig. 10-5d is at the same point B as is the element of Fig. 10-5c, the big question is: What relationship exists between the stress state for these two elements?

To explore the problem as simply as possible, let us select the numerical example of Fig. 10-6a in which $\sigma_x = 10$ psi, $\tau_{xy} = 4$ psi, and the element has

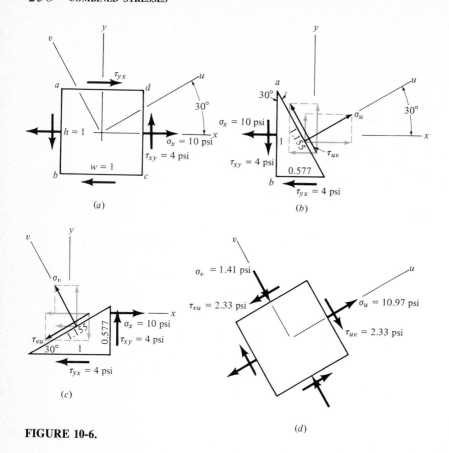

FIGURE 10-6.

sides t, h, and w all 1 in in length. We select a set of uv axes at an angle of $\phi = 30°$ measured ccw from the xy system. The problem now reduces to that of finding the stresses σ_u, σ_v, and τ_{uv}, together with their directions.

Beginning at corner a, in Fig. 10-6b, we cut the element at an angle of 30° in order to produce a face normal to the u axis. This produces a triangular element having a vertical face still 1 in long, a base of length 1 tan 30° = 0.577 in wide, and a hypotenuse of length 1/cos 30° = 1.155 in, as shown. We are now going to apply the principles of statics by summing the forces in the x direction and in the y direction to zero. Force equals the stress on the side of the element times the area of the side. To make the calculations which follow clear, the areas are shown in brackets. Remember, the depth of the element is 1 in. For the x direction, we have

$$\Sigma F_x = -\sigma_x[(1)(1)] - \tau_{yx}[(0.577)(1)] - \tau_{uv}[(1.155)(1)]\sin 30°$$
$$+ \sigma_u[(1.155)(1)]\cos 30° = 0$$

Next, multiply out the areas and substitute $\sigma_x = 10$ psi, $\tau_{yx} = 4$ psi, sin 30° = 0.5, and cos 30° = 0.866 to get

$$\Sigma F_x = -10 - 4(0.577) - \tau_{uv}(1.155)(0.5) + \sigma_u(1.155)(0.866) = 0$$

Multiplying and rearranging finally gives

$$\sigma_u - 0.577\tau_{uv} = 12.31 \qquad (e)$$

Next, we sum the forces in the y direction. This gives

$$\Sigma F_y = -\tau_{xy}[(1)(1)] + \tau_{uv}[(1.155)(1)] \cos 30° + \sigma_u[(1.155)(1)] \sin 30° = 0$$

Following the same procedure used to derive Eq. (e), we get

$$0.577\sigma_u + \tau_{uv} = 4 \qquad (f)$$

You may now solve Eqs. (e) and (f) simultaneously in any way you choose. The answers are

$$\sigma_u = 10.97 \text{ psi} \qquad \tau_{uv} = -2.33 \text{ psi} \qquad (g)$$

Of course, since τ_{uv} is negative, the direction assumed for it in Fig. 10-6b is incorrect.

To obtain the stresses on the v face, proceed in the same manner using the triangular element of Fig. 10-6c. Do it. You will obtain:

$$\sigma_v = -1.41 \text{ psi} \qquad \tau_{vu} = 2.33 \text{ psi} \qquad (h)$$

Note that σ_v is negative and is compression, not tension as assumed. But, τ_{vu} came out correctly because we knew its proper direction from the preceding analysis.

The complete stress element aligned in the uv directions is shown in Fig. 10-6d.

10-5. TRANSFORMATION OF STRESS

In the previous section, the stress state at a point is described by the stress components acting on the sides of a plane stress element. We learned there that any stress state can be described by referring the stress components to an xy reference system, or alternately, to a uv reference system. The act of converting these components from one system to another is called *transformation*. The purpose of this section is to investigate the general relations involved in the transformation of stress from one reference system to another.

We begin with the general plane-stress element of Fig. 10-7a in which the stress components referred to an xy system σ_x, σ_y, τ_{xy}, and τ_{yx}. Of course, $\tau_{yx} = \tau_{xy}$. The problem is to find the stress components on another face whose normal makes an angle of ϕ from the x axis, measured in the ccw direction from x. Accordingly, we construct the triangular stress element in Fig. 10-7b, with the inclined face having its normal at the angle ϕ from x. We choose not to designate this as the u axis, as in the preceding section, and so the normal and shear stresses acting on the inclined face are simply σ and τ, as shown. The vertical, or x face, ab, has a height h. Therefore, the horizontal, or y face, has a length of $h \tan \phi$. And the length of the hypotenuse is $h/\cos \phi$. The depth of the element normal to the paper is taken as one unit of length.

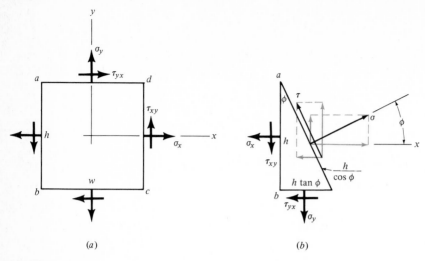

FIGURE 10-7.

Note that the stresses σ and τ acting on the inclined face have been decomposed into horizontal and vertical components. The x and y components of σ are, respectively,

$$\sigma \cos \phi \qquad \sigma \sin \phi$$

Similarly, the x and y components of τ are seen to be

$$-\tau \sin \phi \qquad \tau \cos \phi$$

Our next step is to sum the forces (not stresses) to zero separately in the x and y directions. In doing this, remember, a force is the stress times the area. So, for the x direction, we have

$$\Sigma F_x = -\sigma_x h - \tau_{yx} h \tan \phi + \sigma \cos \phi \frac{h}{\cos \phi} - \tau \sin \phi \frac{h}{\cos \phi} = 0$$

Using the relation $\tan \phi = \sin \phi / \cos \phi$ and rearranging, we obtain

$$\sigma - \tau \tan \phi = \sigma_x + \tau_{xy} \tan \phi \qquad (a)$$

Repeating the same procedure for the y direction gives

$$\Sigma F_y = -\tau_{xy} h - \sigma_y h \tan \phi + \tau \cos \phi \frac{h}{\cos \phi} + \sigma \sin \phi \frac{h}{\cos \phi} = 0$$

which becomes

$$\sigma \tan \phi + \tau = \sigma_y \tan \phi + \tau_{xy} \qquad (b)$$

When we solve Eqs. (a) and (b) simultaneously for σ and τ, we get

$$\sigma = \frac{\sigma_x + 2\tau_{xy} \tan \phi + \sigma_y \tan^2 \phi}{1 + \tan^2 \phi} \qquad (c)$$

$$\tau = \frac{(\sigma_y - \sigma_x)\tan\phi + \tau_{xy}(1 - \tan^2\phi)}{1 + \tan^2\phi} \qquad (d)$$

These equations must now be transformed by introducing the sine and cosine functions of the double angle 2ϕ. Quite a bit of algebra is involved, together with the following trigonometric identities:

$$\tan\phi = \frac{\sin\phi}{\cos\phi} \qquad (e)$$

$$\frac{1}{1 + \tan^2\phi} = \cos^2\phi \qquad (f)$$

$$2\sin\phi\cos\phi = \sin 2\phi \qquad (g)$$

$$\cos^2\phi = \frac{1}{2} + \frac{1}{2}\cos 2\phi \qquad (h)$$

$$\sin^2\phi = \frac{1}{2} - \frac{1}{2}\cos 2\phi \qquad (i)$$

After completing this work, the results can finally be arranged as

$$\sigma = \frac{\sigma_x + \sigma_y}{2} + \frac{\sigma_x - \sigma_y}{2}\cos 2\phi + \tau_{xy}\sin 2\phi \qquad (10\text{-}6)$$

$$\tau = -\frac{\sigma_x - \sigma_y}{2}\sin 2\phi + \tau_{xy}\cos 2\phi \qquad (10\text{-}7)$$

When σ_x, σ_y, and τ_{xy} are given, these equations can be used to find the stress components σ and τ on any face whose normal makes the specified angle ϕ with the x axis. We shall not usually solve such a problem using Eqs. (10-6) and (10-7), however, because a graphical approach to be presented in the next section turns out to be quicker and easier. But, let us next explore some of the characteristics of Eqs. (10-6) and (10-7).

A mathematical technique for determining the maximum and minimum value or values of a function consists in differentiating the function with respect to the independent variable and setting the result equal to zero. When the resulting equation is solved, the result is the value of the independent variable for which the function is a maximum or a minimum. We first learned this in our study of beams, where we learned that the maximum bending moment occurs where the shear force is zero, and that the maximum deflection occurs where the slope is zero.

In our problem here, we find the value of ϕ for which σ is maximum by taking the derivative of Eq. (10-6), setting the result equal to zero, and solving for ϕ. Thus,

$$\frac{d\sigma}{d\phi} = \frac{d}{d\phi}\left(\frac{\sigma_x + \sigma_y}{2}\right) + \frac{d}{d\phi}\left(\frac{\sigma_x - \sigma_y}{2}\cos 2\phi\right) + \frac{d}{d\phi}(\tau_{xy}\sin 2\phi)$$

$$= 0 - (\sigma_x - \sigma_y)\sin 2\phi + 2\tau_{xy}\cos 2\phi = 0 \qquad (j)$$

If we divide this result through by $\cos 2\phi$, rearrange, and transform to $\tan 2\phi$, we get

$$\tan 2\phi = \frac{2\tau_{xy}}{\sigma_x - \sigma_y} \tag{10-8}$$

When this equation is solved, using any set of values of σ_x, σ_y, and τ_{xy}, two values of 2ϕ will result. One of these, when substituted into Eq. (10-6) will give a maximum normal stress, designated as σ_1, and the other value of 2ϕ yields a minimum normal stress, called σ_2. These two stresses are called *principal stresses*, and the corresponding values of ϕ are called the *principal directions*. The principal directions are at right angles to each other.

If, in a similar manner, we take the derivative of Eq. (10-7) with respect to ϕ, set the derivative equal to zero, and solve the result for $\tan 2\phi$, we get

$$\tan 2\phi = -\frac{\sigma_x - \sigma_y}{2\tau_{xy}} \tag{10-9}$$

This equation defines two values of 2ϕ for which the shear stress τ is a maximum.

There is still another interesting characteristic of Eqs. (10-6) and (10-7) to be determined. Suppose we solve Eq. (10-8) for one of the values of 2ϕ. When this value is substituted into Eq. (10-6), we get a principal stress, σ_1 or σ_2. But, when we substitute the same value of 2ϕ into Eq. (10-7), we always get $\tau = 0$. In other words, the shear stress is always zero in the direction of the principal stresses. In fact, this observation is used by some to define the principal directions and the principal stresses. Thus, whenever a stress element is oriented such that the shear stresses are zero, then the resulting normal stresses are the principal stresses and their directions are the principal directions.

To prove that the shear stress associated with both principal stresses is zero, first solve Eq. (j) for $\sin 2\phi$. This gives

$$\sin 2\phi = \frac{2\tau_{xy} \cos 2\phi}{\sigma_x - \sigma_y} \tag{k}$$

Now substitute Eq. (k) for $\sin 2\phi$ in Eq. (10-7); the result is

$$\tau = -\frac{\sigma_x - \sigma_y}{2} \frac{2\tau_{xy} \cos 2\phi}{\sigma_x - \sigma_y} + \tau_{xy} \cos 2\phi = 0 \tag{10-10}$$

In a similar manner, if we solve Eq. (10-9) for $\sin 2\phi$ and substitute the result into Eq. (10-6), we get

$$\sigma = \frac{\sigma_x + \sigma_y}{2} \tag{10-11}$$

which tells us that both normal stresses associated with the directions of the two maximum shear stresses are equal.

Formulas for the two principal stresses can be found by substituting the angle 2ϕ from Eq. (10-8) into Eq. (10-6). Without presenting the required algebra here, the result will be found to be

$$\sigma_1, \sigma_2 = \frac{\sigma_x + \sigma_y}{2} \pm \sqrt{\left(\frac{\sigma_x - \sigma_y}{2}\right)^2 + \tau_{xy}{}^2} \tag{10-12}$$

$$\tau_{max} = \pm \sqrt{\left(\frac{\sigma_x - \sigma_y}{2}\right)^2 + \tau_{xy}{}^2} \tag{10-13}$$

10-6. MOHR'S CIRCLE REPRESENTATION   A graphical method for expressing all the relations derived in the preceding section, called *Mohr's circle diagram,* is a very effective means of visualizing the stress state at a point and keeping track of the directions of the various components of plane stress. In Fig. 10-8, we create a $\sigma\tau$ coordinate system with normal stresses plotted along the abscissa, and shear stresses plotted as the ordinates. On the abscissa, tensile (positive) normal stresses are plotted to the right of the origin O, and compressive (negative) normal stresses are plotted to the left. The sign convention for shear stresses is that clockwise (cw) shear stresses are plotted *above* the abscissa, and counterclockwise (ccw) shear stresses *below*.

Using the stress state of Fig. 10-7a, we plot Mohr's circle diagram (Fig. 10-8) by laying off σ_x as OA, τ_{xy} as AB, σ_y as OC, and τ_{yx} as CD. The line DEB is the diameter of Mohr's circle with center at E on the σ axis. Point B represents the stress coordinates $\sigma_x\tau_{xy}$ on both x faces, and point D the stress coordinates $\sigma_y\tau_{yx}$ on both y faces. Thus, EB corresponds to the x axis and ED to the

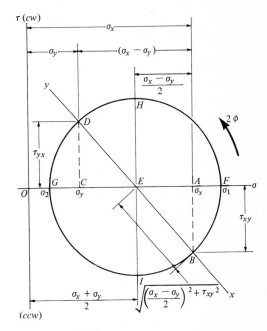

FIGURE 10-8.

y axis. Double angles 2ϕ are measured on the diagram in the same direction, cw or ccw, as the single angles ϕ are measured on the corresponding stress element. Thus, the angle from EB (on the circle) around to ED is 180°. But, the angle from x to y on the stress element is 90°, which is half of 180°.

The maximum principal normal stress σ_1 occurs at F, and the minimum principal normal stress σ_2 occurs at G. The double angle 2ϕ from EB to EF corresponds to the principal direction for σ_1. And the angle from EB to EG, measured in either the cw or the ccw direction, corresponds to the principal direction for σ_2.

EXAMPLE 10-1 Draw Mohr's circle to scale corresponding to the stress state $\sigma_x = 12$ kpsi, $\sigma_y = 0$ kpsi, and $\tau_{xy} = 7$ kpsi cw. Find the principal stresses, principal directions, and the maximum shear stress from the circle by measurement. Draw the principal stress element properly oriented from the xy system and label it completely. Draw another stress element properly oriented to show the maximum shear stresses, find the corresponding normal stresses, and label it completely.

SOLUTION

In Fig. 10-9a, begin the diagram by laying off $\sigma_x = 12$ kpsi to scale along the σ axis. Then, from σ_x, lay off $\tau_{xy} = 7$ kpsi in the cw direction of the τ axis to establish point A. Thus, the coordinates of A are $\sigma_x \tau_{xy}$, and so A represents the x face of the stress element. Next, corresponding to $\sigma_y = 0$ at the origin O, lay off $\tau_{yx} = 7$ kpsi in the ccw direction along the τ axis to establish point B. The coordinates of B are $\sigma_y \tau_{yx}$, which correspond to the y face of the stress element. The line AB forms the diameter of Mohr's circle, which can now be drawn. Note that the angle from A around the circle to B is 180°. Since double angles are measured on the circle, this corresponds to the right angle on the stress element between the x and y axes.

The intersection of the circle with the σ axis locates the two principal stresses: σ_1 at E and σ_2 at F. The values are obtained by scaling the diagram. Note that F is to the left of the origin O, and hence that σ_2 is negative. It would therefore have been unnecessary to add that $\sigma_2 = -3.22$ kpsi to the diagram; this was done for emphasis. The double angle $2\phi = 49.4°$ from A to E is measured in the cw direction on the circle diagram. On the principal stress element (Fig. 10-9b), the corresponding single angle $\phi = 49.4/2 = 24.7°$ is measured in the identical cw direction from the x axis to the first principal stress axis.

Remember, angles which are measured in the cw direction on Mohr's circle must also be measured in the cw direction on the corresponding stress element. Only by following this rule can we be assured that the stress elements are properly oriented. The principal stress element of Fig. 10-9b can now be completed and labeled, as shown.

The two maximum shear stresses, one cw and one ccw, occur at points C and D in Fig. 10-9a. The two normal stresses occurring on the same faces as these are each 6 kpsi, as shown. Point C is oriented from point A by the ccw double angle of

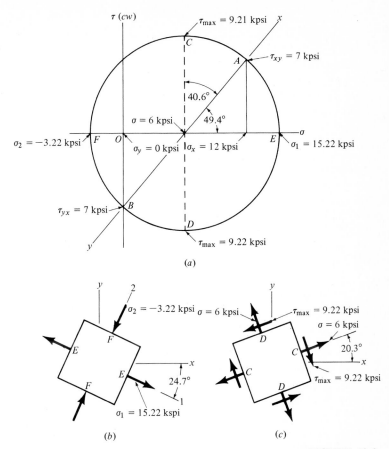

FIGURE 10-9.

40.6° on Mohr's circle. Therefore, in Fig. 10-9c, draw a stress element having the face C oriented 20.3° (half of 40.6°) from x. The element is then labeled by designating the proper magnitude and direction of each stress component, as illustrated.

////

EXAMPLE 10-2   Solve Example 10-1 using Mohr's circle and an analytical approach.

SOLUTION

It is easier to use Mohr's circle with an analytical approach than it is to use the equations developed in the preceding section to solve combined stress problems. The procedure is to make an approximate sketch of Mohr's circle, labeling it with the given data, and then to use the geometry of the figure to obtain the desired information.

Begin the circle diagram of Fig. 10-10 by sketching the circle and labeling it with the x and y directions. Now, noting the triangle ABC, indicate on the sketch the length of the legs AB and BC as 7 and 6, respectively. Note that all stresses are

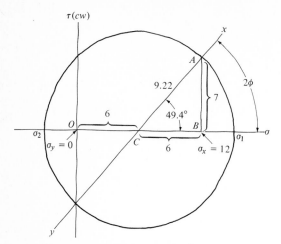

FIGURE 10-10. Mohr's circle drawn freehand; all stresses are in kpsi.

in kpsi in this example. The hypotenuse AC of the triangle is the maximum shear stress, since it is the radius of the circle. It is

$$\tau_{max} = \sqrt{(7)^2 + (6)^2} = 9.22$$

As can be seen from the geometry, the principal stresses are the distances OC plus CA and OC minus CA. Therefore,

$$\sigma_1 = 6 + 9.22 = 15.22$$
$$\sigma_2 = 6 - 9.22 = -3.22$$

Finally, from triangle ABC, the double angle from the x axis to the σ_1 axis on Mohr's circle is

$$2\phi = \tan^{-1} \frac{7}{6} = 49.4°$$

Therefore, the angle from x to the σ_1 axis on the stress element is half of 49.4° (24.7°) in the cw direction.   ////

10-7. APPLICATIONS   When a member of a structure or a machine is subjected to a complex load or loads, the analysis problem is usually to find the principal stresses, the maximum shear stress, and the directions. This information would then be used to find out if a part is strong enough to carry the load. In some cases, the results of the analysis might be used to explain why a member had failed in use.

Figure 10-11a shows a pulley shaft that is a case of combined torsion and bending. The belt pulls cause the shaft to bend in the xz plane, with the positive z side of the shaft being in tension. Also, the pulleys subject the shaft to torque because one pulley, say, may receive power from a motor, or other source, and transmit this through the shaft to the other pulley, which then transmits it

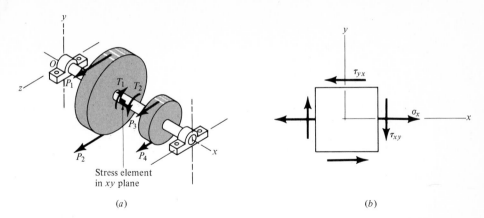

Stress element
in xy plane

(a)

(b)

through belts to a machine on the receiving end. From our previous studies, we know the stresses to be

$$\sigma_x = \frac{Mc}{I} \qquad (a)$$

$$\tau_{xy} = \frac{Tr}{J} \qquad (b)$$

One would normally calculate these stresses at a place on the shaft where the bending moment is maximum. If the shaft were shouldered, however, with various sections having different diameters, then the maximum stress might occur at a place other than the location of the maximum bending moment. Once the component stresses have been computed for the significant stress element (Fig. 10-11b), Mohr's circle can be used to obtain the principal stresses and directions and the maximum shear stress.

Figure 10-12a shows a short rectangular cantilever carrying a uniformly distributed bending load. The stress elements A, B, and C in Fig. 10-12b show how bending stress may occur in combination with direct shear stress. For a stress element at A on the top surface of the cantilever, the shear stress is zero and the bending stress is given by Eq. (a). However, an element at B, a distance y from the neutral axis, has a bending stress of

$$\sigma_x = \frac{My}{I} \qquad (c)$$

The direct shear stress, from Eq. (5-10), is

$$\tau_{xy} = \frac{3V}{2A}\left(1 - \frac{y^2}{c^2}\right) \qquad (d)$$

Finally, the element at C is on the neutral axis, and so $y = 0$ and $\sigma_x = 0$. The direct shear stress reaches a maximum of

$$\tau_{xy} = \frac{3V}{2A} \qquad (e)$$

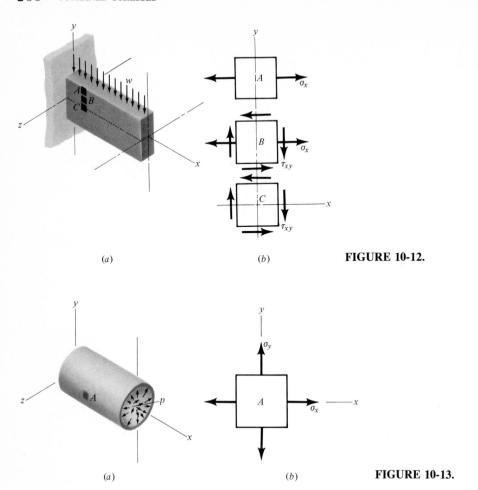

(a)   (b)   **FIGURE 10-12.**

(a)   (b)   **FIGURE 10-13.**

A third example of the need for studying combined stress is furnished by the section of a thin-wall pressure cylinder in Fig. 10-13a. Due to an internal pressure p, we learned in Sec. 3-8 that there exists in the shell a longitudinal stress of magnitude

$$\sigma_x = \frac{pd_i}{4t} \qquad (f)$$

where t is the thickness, p the pressure, and d_i the inside diameter. There also exists a hoop or tangential stress that is normal to the longitudinal component. If we choose the element A in Fig. 10-13, then the hoop stress is in the y direction, and we have

$$\sigma_y = \frac{pd_i}{2t} \qquad (g)$$

Since there are no shear stresses acting, these are the principal stresses.

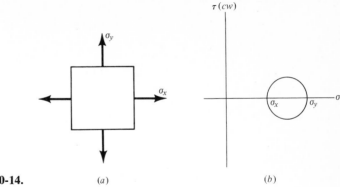

FIGURE 10-14.   (a)   (b)

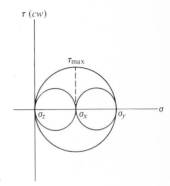

FIGURE 10-15.

The stress element corresponding to Eqs. (*f*) and (*g*) is shown in Fig. 10-14*a*, and the accompanying Mohr's circle in part *b*. There is an opportunity of making a very serious error in analysis when both normal stress components have the same sign. This is illustrated by Fig. 10-14*b*. In computing the maximum shear stress, there is a danger of computing it as the radius of the circle that passes through σ_x and σ_y. But, since $\sigma_z = 0$, another larger circle can be drawn through σ_y and σ_z, and the maximum shear stress is then found to be

$$\tau_{max} = \frac{\sigma_y}{2} \tag{h}$$

This is shown by the diagram of Fig. 10-15. Note that three circles can be drawn, one through σ_y and σ_x, another through σ_y and σ_z, and the third through σ_x and σ_z. This is sometimes called *Mohr's three-circle diagram*.

PROBLEMS

SECTION 10-2

10-1 The horizontal member OBC in the figure has a force $F = 2$ lb acting at C. Compute the stresses on the top and bottom of OBC at a section $1\frac{1}{4}$ in from O.

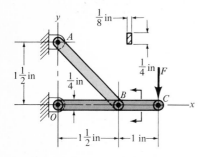

PROB. 10-1.

10-2 A force $F = 3200$ lb is applied in the horizontal direction at C on forging OBC in the figure. Due to this force, the maximum stress would normally occur at B. However, a reinforcement around B extends to a distance of $1\frac{1}{2}$ in on both sides. Therefore, compute the maximum tensile stress and the maximum compressive stress in OBC at a distance of $1\frac{1}{2}$ in from B. On which side of B do these occur?

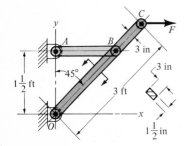

PROB. 10-2.

10-3 Find the maximum tensile and compressive stresses on section Q of the weldment ABC illustrated in the figure. The bar is loaded at C by a vertical force $F = 2400$ lb.

10-4 Determine the stress distribution at a section midway between B and D of the wood beam BDE in the figure. Applied at E is a force $F = 1600$ lb.

10-5 The figure shows a wood A frame loaded by a force $F = 800$ lb acting at the apex. Compute the maximum tensile and compressive stresses on the 4×10-in timber CDE on a section at Q, as shown.

10-6 A force $F = 4200$ lb is applied to the aluminum-alloy L bar shown in the figure. Find the maximum tensile and compressive stresses on a section at Q.

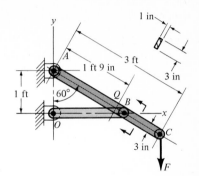

PROB. 10-3.

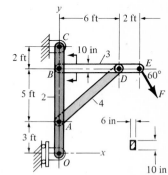

PROB. 10-4.

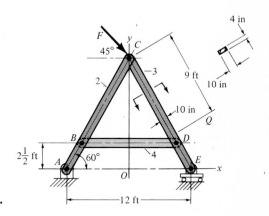

PROB. 10-5.

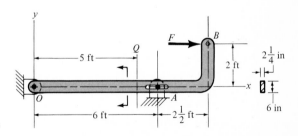

PROB. 10-6.

10-7   A force $F = 900$ lb acts at point B on the overhanging L bar shown in the figure. Compute the maximum tensile and compressive stresses acting on sections at P and Q.

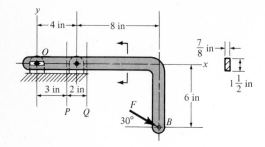

PROB. 10-7.

10-8   The figure illustrates a table-model pneumatic impact press. A power cylinder mounted in guide bracket A may slide vertically on a base-supported column C. In use, a chuck on piston rod D holds a tool, which may be used for such purposes as riveting, swaging, or staking, for example. The force of the impact is adjustable up to 700 lb. Based on this force, find the maximum stress in column C.

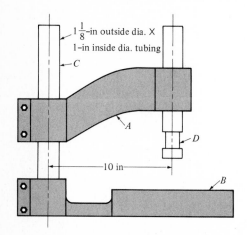

PROB. 10-8.

10-9   The C clamp shown in the figure has a body made with a trapezoidal cross section, as shown in one of the sections of the figure. Find the maximum and minimum stress on a section at A-A if the clamping force is 800 lb.

10-10   The same as Prob. 10-9, except use the rectangular cross section.

10-11   The same as Prob. 10-9, except use the T cross section.

10-12   Find the stresses in the frame at section A-A of the coping saw shown in the figure. Jaw B, which holds one end of the blade, has a threaded shank that screws into the handle of the saw. When the handle is tightened, the saw blade is under a tension of 15 lb.

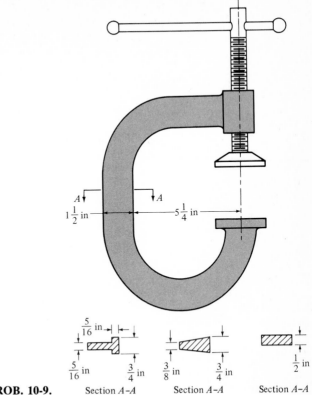

PROB. 10-9.   Section *A–A*   Section *A–A*   Section *A–A*

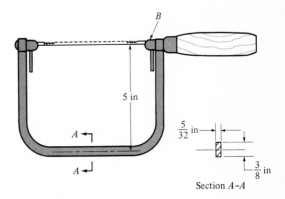

PROB. 10-12.   Section *A–A*

10-13 The blade of the bow saw shown in the figure is tightened by a wing nut, which then results in a maximum blade tension of 120 lb. Find the resulting stresses in the tubular frame at section *Q*.

10-14 The press frame shown in the figure is planned for casting using ASTM No. 25 gray cast iron. Based upon stress calculations made at section *A-A*, would the press frame be more likely to fail in tension or in compression?

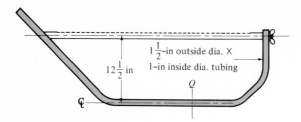

PROB. 10-13.

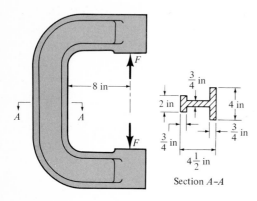

Section *A–A*

PROB. 10-14.

SECTION 10-4

10-15 A plane stress element has $\sigma_x = 100$ psi, $\sigma_y = -300$ psi, and $\tau_{xy} = 0$. What would be the stresses on the same element if it were rotated 45° in the ccw direction from its present position?

10-16 A plane stress element has both normal stresses zero, but the shear stress is $\tau_{xy} = 32$ psi cw. What would be the stresses on the element if it were rotated 60° in the ccw direction from its present orientation?

10-17 The stresses acting at a point on a member are $\sigma_x = 0$, $\sigma_y = -400$ psi, and $\tau_{xy} = 200$ psi cw. Find the stresses on an element at the same point having faces normal to the *uv* axes. The *u* axis is oriented 30° in the ccw position from *x*.

10-18 The stresses acting at a point on a member are $\sigma_x = -100$ psi, $\sigma_y = 200$ psi, and $\tau_{xy} = 120$ psi ccw. A stress element is to be defined at this point, whose faces are normal to a *uv* system in which the *u* axis is oriented 30° in the cw direction from the *x* axis. Find the stresses and their directions corresponding to this element.

SECTION 10-5[1]

10-19 At what angles measured from the *x* axis do the principal stresses occur if $\sigma_x = 10$ kpsi, $\sigma_y = 0$, and $\tau_{xy} = 8$ kpsi ccw? Compute the values of the principal stresses.

---

[1] For this group of problems only, treat the shear stress as negative if in the clockwise direction.

10-20   A stress element has $\sigma_x = 0$, $\sigma_y = -12$ kpsi, and $\tau_{xy} = 6$ kpsi cw. Find the principal directions and principal stresses.

10-21   Find the principal directions and principal stresses if $\sigma_x = 6$ kpsi, $\sigma_y = -6$ kpsi, and $\tau_{xy} = 6$ kpsi cw.

10-22   Find the principal directions and stresses if $\sigma_x = 18$ kpsi, $\sigma_y = -4$ kpsi, and $\tau_{xy} = 10$ kpsi ccw.

10-23   For each of the following stress states, construct Mohr's circle and find the principal stresses and the maximum shear stresses. Draw a stress element properly oriented from the x axis to show the principal stresses and their directions. Calculate the angle between the x axis and σ_1.
   (a) $\sigma_x = 10$ kpsi, $\sigma_y = 0$ kpsi, $\tau_{xy} = 4$ kpsi cw
   (b) $\sigma_x = -2$ kpsi, $\sigma_y = -8$ kpsi, $\tau_{xy} = 4$ kpsi ccw
   (c) $\sigma_x = 10$ kpsi, $\sigma_y = 5$ kpsi, $\tau_{xy} = 6$ kpsi cw

10-24   The same as Prob. 10-23, but using the following stress states:
   (a) $\sigma_x = 0$ kpsi, $\sigma_y = -22$ kpsi, $\tau_{xy} = 12$ kpsi cw
   (b) $\sigma_x = 4$ kpsi, $\sigma_y = 24$ kpsi, $\tau_{xy} = 12$ kpsi ccw
   (c) $\sigma_x = -4$ kpsi, $\sigma_y = 18$ kpsi, $\tau_{xy} = 8$ kpsi ccw

10-25   The same as Prob. 10-23, but using the following stress states:
   (a) $\sigma_x = -2$ kpsi, $\sigma_y = -16$ kpsi, $\tau_{xy} = 10$ kpsi ccw
   (b) $\sigma_x = 36$ kpsi, $\sigma_y = 0$ kpsi, $\tau_{xy} = 18$ kpsi ccw
   (c) $\sigma_x = 0$ kpsi, $\sigma_y = 32$ kpsi, $\tau_{xy} = 12$ kpsi cw

10-26   The same as Prob. 10-23, except use the following stress states:
   (a) $\sigma_x = 14$ kpsi, $\sigma_y = 14$ kpsi, $\tau_{xy} = 14$ kpsi cw
   (b) $\sigma_x = 12$ kpsi, $\sigma_y = -12$ kpsi, $\tau_{xy} = 6$ kpsi ccw
   (c) $\sigma_x = 6$ kpsi, $\sigma_y = -18$ kpsi, $\tau_{xy} = 12$ kpsi cw

10-27   A plane stress element has both normal stresses zero, but the shear stress is $\tau_{xy} = 32$ kpsi cw.
   (a) Compute the principal stresses and principal directions.
   (b) Compute the stresses on an element that has been rotated 60° ccw from the x axis.

10-28   A plane stress element has $\sigma_x = 800$ psi, $\sigma_y = -400$ psi, and $\tau_{xy} = 800$ psi ccw. Calculate the stresses on another element having a u axis 30° cw from the x axis. Draw the element and show all stresses which act.

10-29   A plane stress element has $\sigma_x = 18$ kpsi, $\sigma_y = -6$ kpsi, and $\tau_{xy} = 0$ kpsi. Find the stress components acting on another element at the same point having a u axis 75° ccw from the x axis.

10-30   If $\sigma_x = -12$ kpsi, $\sigma_y = 4$ kpsi, and $\tau_{xy} = 8$ kpsi cw, what are the stresses on another element at the same point having a u axis 45° cw from the x axis?

10-31   The figure shows a countershaft subjected to bending and torsion by the two pulleys. For each pulley, the belt tensions are applied parallel to the xz plane. A stress element at point B, midway between the two pulleys, is on the z side of the shaft and hence has normals parallel to x and y.

(a) Find the stress components which act upon this element.

(b) Find the corresponding principal stresses and their orientation from the x axis.

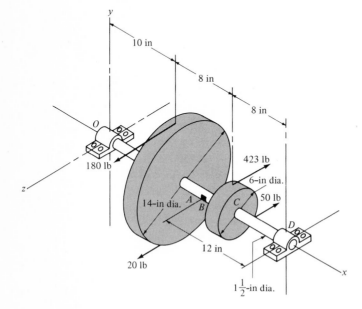

PROB. 10-31.

10-32 The $1\frac{1}{4}$-in diameter steel-pulley shaft shown supports two inboard pulleys at A and B and an outboard (or overhanging pulley) at D.

(a) Find the shaft torque between A and B and between B and D.

(b) Find the belt tensions P_1 and P_2 if the tension on the loose side is 10 percent of that on the tight side.

(c) Locate a stress element on the shaft midway between the pulley at B and the bearing at C and on the side of the shaft subjected to a tensile bending stress such that the sides are parallel to the reference system. Find the stress components which act upon this element.

(d) Find the principal stresses for the element of part c and their orientation with respect to the xyz system of axes.

10-33 A force of 700 lb acts downward on bar CD of the figure, producing twisting and bending in OC. Shown near the origin are the stress element A, having sides normal to the x and z axes, and element B, with sides normal to the x and y axes.

(a) Find all the stress components which act on each element and make sketches of the elements to show the directions in which these act.

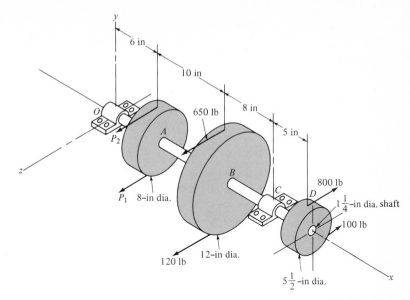

PROB. 10-32.

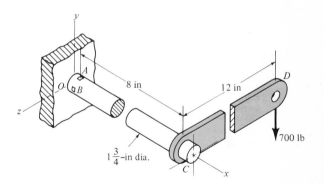

PROB. 10-33.

(b) Using Mohr's circle, find the principal stresses and the principal directions. Make sketches of the principal stress elements and show their orientation with respect to the *xyz* reference system.

10-34 The figure shows the crankshaft of a one-cylinder air compressor. A torque $T = 5400$ lb · in is instantaneously applied from the flywheel to the crank to obtain the piston force P. A stress element is located on the top surface of the crankshaft at A, 4 in from the bearing, and with sides parallel to the x and z axes.

(a) Compute the stress components that act at A.

(b) Find the principal stresses and their directions for point A. Draw the principal stress element and show how it is aligned with respect to the *xz* axes.

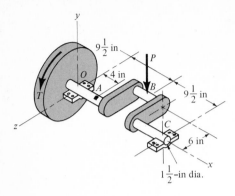

PROB. 10-34.

SI UNITS

10-35 In the figure for Prob. 10-33, let $AC = 200$ mm and $CD = 300$ mm. The shaft diameter is 40 mm and the force at D is 3 kN. The two stress elements at A and B have normals parallel to the y and z axes, respectively.

(a) Find all the stress components that act on each element and make appropriate sketches to show the directions in which these act.

(b) Using Mohr's circle, find the principal stresses and the corresponding directions. Sketch the principal stress elements and show their orientation with respect to the coordinate system.

stress–strength relations

11-1. INTRODUCTION One of the primary considerations in building any structure or machine is that the strength must be sufficiently greater than the stress to assure an adequate margin of safety. The problem is not simple. In Chap. 10, we learned that there can exist a number of stress components at a point in a loaded member. We also learned that these stresses varied in some manner as the location of the point is moved from place to place on the member. It is also true that the strength may vary from point to point in a member due to the method of processing or of forming the part. This subject, however, is too lengthy to be considered in this book. The important consideration is that we have several stress components, but only one strength, usually the yield strength or the ultimate strength. How can we determine the margin of safety by relating several stress components to a single value of strength? To answer this question, we shall study some of the currently used theories of why parts fail.

The other problem to be studied in this chapter is the relation between strength, and stresses which are rapidly reversed thousands and thousands of times during the life of the part. Such stresses are called *fatigue stresses,* and we shall find that parts subjected to them may have corresponding strengths significantly lower than the yield strength of the material.

11-2. THE MAXIMUM NORMAL STRESS THEORY When various stress components act at a point in a body, it is always possible to use Mohr's circle methods to obtain the principal stresses. In studying failure theories, we shall assume that this has been done. The resulting stress state at a point can then be diagrammed as in Fig. 11-1. The three principal stresses σ_1, σ_2, and σ_3 are all shown, with the third, σ_3, shown to emphasize that it must not be forgotten. [See Sec. 10-7, Eq. (h).] A plane stress state exists when one of these principal stresses is zero; such situations are most common.

No matter what values the principal stresses have, the subscripts can always be interchanged. Thus, we shall assume that $\sigma_1 > \sigma_2 > \sigma_3$ in all cases, with σ_1 being the greatest positive stress and σ_3 being the least (or most negative) stress.

The maximum normal stress theory states that *failure occurs when the largest principal normal stress equals the strength.* All other principal stresses are ignored in using the theory. And either the yield strength or the ultimate strength can be used. Thus, for tension, failure occurs whenever

$$\sigma_1 = S_{yt} \quad \text{or} \quad \sigma_1 = S_{ut} \tag{a}$$

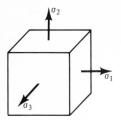

FIGURE 11-1.

And a pure compressive failure would occur when

$$\sigma_3 = -S_{yc} \quad \text{or} \quad \sigma_3 = -S_{uc} \quad\quad (b)$$

provided, of course, that σ_1 is a tensile stress and σ_3 a compressive stress.

Sometimes, it is desirable to define safety instead of failure. In that case, Eqs. (a) and (b) would become

$$\sigma_1 = \frac{S_{yt}}{n} \quad \text{or} \quad \sigma_1 = \frac{S_{ut}}{n} \quad\quad (c)$$

and

$$\sigma_3 = -\frac{S_{yc}}{n} \quad \text{or} \quad \sigma_3 = -\frac{S_{uc}}{n} \quad\quad (d)$$

where n is the factor of safety.

The maximum normal stress theory is an easy one to use, but it may give results that are on the unsafe side. In fact, it predicts that the yield strength in shear is the same as the yield strength in tension, which is contrary to experimental evidence.

11-3. THE MAXIMUM SHEAR STRESS THEORY   This is an easy theory to use; it is always on the safe side of test results and it is in use in many construction and safety codes. Since it is used only to predict when yielding occurs, it applies only to ductile materials, that is, materials which have a yield strength.

Figure 11-2 shows Mohr's circle diagram for the simple tension test. For this test, σ_2 and σ_3 are both zero and yielding occurs when $\sigma_1 = S_{yt}$, as shown. For this circle, the maximum shear stress is half the yield strength. This is the basis of the theory. The maximum shear stress theory states that *failure occurs*

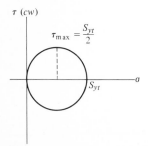

FIGURE 11-2.

whenever the maximum shear stress at a point equals the maximum shear stress in the tension-test specimen when that specimen begins to yield.

Thus, for a general stress state, failure occurs whenever

$$\frac{\sigma_1 - \sigma_3}{2} = \frac{S_{yt}}{2} \qquad (a)$$

because, with $\sigma_1 > \sigma_2 > \sigma_3$, the maximum shear stress is $(\sigma_1 - \sigma_3)/2$. Safety occurs when

$$\frac{\sigma_1 - \sigma_3}{2} = \frac{S_{yt}}{2n} \qquad (b)$$

where, again, n is the factor of safety.

According to this theory, the yield strength in shear is

$$S_{sy} = \frac{S_{yt}}{2} \qquad (c)$$

All three equations in this section can also be written in terms of the yield strength in compression, S_{yc}.

11-4. THE DISTORTION-ENERGY THEORY   This is a difficult theory to derive, but it is an important one because it comes closest of all to verifying experimental results. The derivation is based on the assumption that Hooke's law (*stress is proportional to strain*) applies, and so this theory is valid only for ductile materials, that is, materials which may yield.

In the literature, you may find the distortion-energy theory also called the *shear-energy theory,* as well as the *von Mises-Hencky theory.*

The distortion-energy theory originated because of the observation that ductile materials stressed hydrostatically (equal tension or equal compression along three coordinate axis) had yield strengths greatly in excess of the values given by the simple tension test. It was postulated that yielding was not a simple tensile or compressive phenomenon at all, but rather, that it was related some-how to the angular distortion of the stressed element. One of the earlier theories of failure, now abandoned, predicted that yielding would begin whenever the total energy stored in the stressed element became equal to the energy stored in an element of the tension-test specimen at the yield point. This theory was called the *maximum strain energy theory;* it was a forerunner of the distortion-energy theory. The distortion-energy theory is based on subtracting from the total energy that which produces only a simple volume change. Then, the energy left produces only angular distortion. We compare this energy with the distortion energy in an element of the tension-test specimen to predict yielding.

Figure 11-3 shows an element having sides of length w, h, and t, acted upon by stresses arranged so that $\sigma_1 > \sigma_2 > \sigma_3$. The stress σ_1 produces a force resultant on the side of the element of magnitude $F = \sigma_1 A = \sigma_1(ht)$. This same force produces an elongation of the element of magnitude $\delta_1 = \epsilon_1 w$. The energy stored

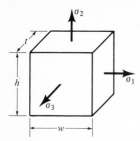

FIGURE 11-3.

by applying this stress, called *strain energy,* is half the force times the elongation, or

$$U_1 = \frac{F\delta_1}{2} = \frac{\sigma_1(ht)(\epsilon_1 w)}{2}$$ (*a*)

By using a unit element $h = t = w = 1$ and Eq. (*a*), we get

$$U_1 = \frac{\sigma_1 \epsilon_1}{2}$$ (*b*)

Similarly, for the other two directions, we have

$$U_2 = \frac{\sigma_2 \epsilon_2}{2} \qquad U_3 = \frac{\sigma_3 \epsilon_3}{2}$$ (*c*)

The total strain energy stored in the element is

$$U = U_1 + U_2 + U_3$$ (*d*)

In order to obtain the total strain energy in terms of the stresses, we must first
relate the stresses to the strains. The transformation is something like Eq. (3-22),
but extended to the case of a three-dimensional stress element. The relationship
for all three strains is

$$\epsilon_1 = \frac{\sigma_1}{E} - \frac{\mu\sigma_2}{E} - \frac{\mu\sigma_3}{E}$$

$$\epsilon_2 = \frac{\sigma_2}{E} - \frac{\mu\sigma_3}{E} - \frac{\mu\sigma_1}{E}$$ (*e*)

$$\epsilon_3 = \frac{\sigma_3}{E} - \frac{\mu\sigma_1}{E} - \frac{\mu\sigma_2}{E}$$

We shall omit the algebra here. However, if we substitute Eqs. (*b*) and (*c*) into
(*d*) and then eliminate ϵ_1, ϵ_2, and ϵ_3 with Eqs. (*e*), we get

$$U = \frac{1}{2E} \left[\sigma_1{}^2 + \sigma_2{}^2 + \sigma_3{}^2 - 2\mu(\sigma_1\sigma_2 + \sigma_2\sigma_3 + \sigma_3\sigma_1) \right]$$ (*f*)

This is the total strain energy. We want to subtract from this the strain
energy that produces pure volume change. A stress element subjected to three
equal normal stresses would undergo only volume change. So we define a stress

$$\sigma_{av} = \frac{\sigma_1 + \sigma_2 + \sigma_3}{3}$$ (*g*)

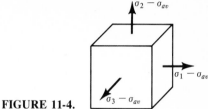

FIGURE 11-4.

and subtract it from σ_1, σ_2, and σ_3. This produces the stress element of Fig. 11-4, which undergoes angular distortion with no volume change. The amount of strain energy U_v that *does* produce volume change can be found by substituting σ_{av} for each principal stress in Eq. (f). The result, after rearrangement, is

$$U_v = \frac{3\sigma_{av}^2}{2E} (1 - 2\mu) \qquad (h)$$

Now, substitute Eq. (g) into (h) and simplify the result. You should then obtain

$$U_v = \frac{1 - 2\mu}{6E} [\sigma_1^2 + \sigma_2^2 + \sigma_3^2 + 2(\sigma_1\sigma_2 + \sigma_2\sigma_3 + \sigma_3\sigma_1)] \qquad (i)$$

Then, the distortion energy U_d is found by subtracting Eq. (i) from (f). This gives

$$U_d = U - U_v = \frac{1 + \mu}{3E} \left[\frac{(\sigma_1 - \sigma_2)^2 + (\sigma_2 - \sigma_3)^2 + (\sigma_3 - \sigma_1)^2}{2} \right] \qquad (11\text{-}1)$$

It is readily seen from this equation that $U_d = 0$ if $\sigma_1 = \sigma_2 = \sigma_3$, as it should be.

We must next use Eq. (11-1) to get the distortion energy stored in an element of the tension-test specimen when it begins to yield. At this stress level, $\sigma_1 = S_y$ and $\sigma_2 = \sigma_3 = 0$. Therefore, Eq. (11-1) gives

$$U_d = \frac{1 + \mu}{3E} S_y^2 \qquad (11\text{-}2)$$

The distortion-energy theory states that yielding will begin for any stress element when the distortion energy is the same as that in an element of the tension-test specimen when yielding begins. If we therefore place Eqs. (11-1) and (11-2) equal to each other, we obtain a relation that states how large the three principal stresses are at the onset of yielding. This gives

$$(\sigma_1 - \sigma_2)^2 + (\sigma_2 - \sigma_3)^2 + (\sigma_3 - \sigma_1)^2 = 2S_y^2 \qquad (11\text{-}3)$$

This is really the complete answer to our problem. These are the values of σ_1, σ_2, and σ_3 that will cause yielding to begin. However, this is for a triaxial stress state. We are mostly concerned with biaxial or plane stresses. For these, $\sigma_2 = 0$ and Eq. (11-3) reduces to

$$\sigma_1^2 - \sigma_1\sigma_3 + \sigma_3^2 = S_y^2 \qquad (11\text{-}4)$$

For analytical purposes, it is convenient to define a *von Mises stress* as

$$\sigma' = \sqrt{\sigma_1^2 - \sigma_1\sigma_3 + \sigma_3^2} \qquad (11\text{-}5)$$

Then, failure is predicted when

$$\sigma' = S_y \tag{j}$$

and safety is predicted by

$$\sigma' = \frac{S_y}{n} \tag{k}$$

For pure torsion, $\sigma_1 = \tau$ and $\sigma_3 = -\sigma_1$. Substituting these into Eq. (11-4) shows that failure will occur in torsion when

$$\tau = \frac{S_y}{\sqrt{3}}$$

Since $1/\sqrt{3} = 0.577$, this means that the yield strength in shear, as predicted by the distortion-energy theory, is

$$S_{sy} = 0.577S_y \tag{k}$$

EXAMPLE 11-1   A ductile steel has a yield strength of 40 kpsi. Find factors of safety corresponding to failure by the maximum normal stress theory, the maximum shear stress theory, and the distortion-energy theory, for each of the following stress states:
(a) $\sigma_x = 10$ kpsi, $\sigma_y = -4$ kpsi
(b) $\sigma_x = 10$ kpsi, $\tau_{xy} = 4$ kpsi cw
(c) $\sigma_x = -2$ kpsi, $\sigma_y = -8$ kpsi, $\tau_{xy} = 4$ kpsi ccw
(d) $\sigma_x = 10$ kpsi, $\sigma_y = 5$ kpsi, $\tau_{xy} = 1$ kpsi cw

SOLUTION
(a) A Mohr's circle diagram will reveal that $\sigma_1 = 10$ kpsi, $\sigma_3 = -4$kpsi, and $\tau_{max} = 7$ kpsi. For the maximum normal stress theory,

$$n = \frac{S_{yt}}{\sigma_1} = \frac{40}{10} = 4 \quad Ans.$$

For the maximum shear stress theory, we have

$$n = \frac{S_{yt}/2}{\tau_{max}} = \frac{40/2}{7} = 2.86 \quad Ans.$$

To use the distortion-energy theory, we first compute the von Mises stress using Eq. (11-5). This gives

$$\sigma' = \sqrt{\sigma_1^2 - \sigma_1\sigma_3 + \sigma_3^2} = \sqrt{(10)^2 - (10)(-4) + (-4)^2} = 12.5 \text{ kpsi}$$

Then the factor of safety is

$$n = \frac{S_{yt}}{\sigma'} = \frac{40}{12.5} = 3.20 \quad Ans.$$

(b) First, using Mohr's circle, find $\sigma_1 = 11.4$ kpsi, $\sigma_3 = -1.4$ kpsi, and $\tau_{max} = 6.4$ kpsi. Then, the factor of safety, using the maximum normal stress theory, is

$$n = \frac{S_{yt}}{\sigma_1} = \frac{40}{11.4} = 3.51 \qquad Ans.$$

For the maximum shear stress theory, we have

$$n = \frac{S_{yt}/2}{\tau_{max}} = \frac{40/2}{6.4} = 3.12 \qquad Ans.$$

The von Mises stress is

$$\sigma' = \sqrt{\sigma_1{}^2 - \sigma_1\sigma_3 + \sigma_3{}^2} = \sqrt{(11.4)^2 - (11.4)(-1.4) + (-1.4)^2} = 12.2 \text{ kpsi}$$

Therefore, the factor of safety, using the distortion-energy theory, is

$$n = \frac{S_{yt}}{\sigma'} = \frac{40}{12.2} = 3.28 \qquad Ans.$$

(c) Again we use a Mohr's circle to find that $\sigma_1 = 0$ kpsi, $\sigma_3 = -10$ kpsi, and $\tau_{max} = 5$ kpsi. This stress situation doesn't require a failure theory because it corresponds one-to-one with the pure compression test; by definition, therefore, all three theories must give the same factor of safety. The factor of safety is

$$n = \frac{-S_{yc}}{\sigma_3} = \frac{-40}{-10} = 4 \qquad Ans.$$

(d) For this stress state, a Mohr's circle diagram yields $\sigma_1 = 10.19$ kpsi, $\sigma_2 = 4.81$ kpsi, and $\sigma_3 = 0$ kpsi. Therefore,

$$\tau_{max} = \frac{\sigma_1}{2} = \frac{10.19}{2} = 5.1 \text{ kpsi}$$

Then, using the maximum normal stress theory, we find the factor of safety to be

$$n = \frac{S_{yt}}{\sigma_1} = \frac{40}{10.19} = 3.93 \qquad Ans.$$

For the maximum shear stress theory, we have

$$n = \frac{S_{yt}/2}{\tau_{max}} = \frac{40/2}{5.1} = 3.93 \qquad Ans.$$

For this stress state, the von Mises stress must be computed using σ_1 and σ_2 because $\sigma_3 = 0$. Thus,

$$\sigma' = \sqrt{\sigma_1{}^2 - \sigma_1\sigma_2 + \sigma_2{}^2} = \sqrt{(10.19)^2 - (10.19)(4.81) + (4.81)^2} = 8.85 \text{ kpsi}$$

Thus, the factor of safety is

$$n = \frac{S_{yt}}{\sigma'} = \frac{40}{8.85} = 4.52 \qquad Ans.$$

It is worth observing now that in parts **a** and **b**, σ_1 and σ_3 had opposite signs. For both of these cases, the largest factor of safety was obtained using the maximum normal stress theory, and the smallest by the maximum shear stress theory. And in part **d**, both principal stresses have the same sign. For this stress state, the distortion-energy theory gave the largest factor of safety, while the remaining two theories gave the same result. The reasons for this behavior will become apparent in the section that follows.   ////

11-5. FAILURE THEORIES FOR DUCTILE MATERIALS   A ductile material is usually classified as a material that has a yield strength and that exhibits more than 5 percent elongation in the standard tension test. Such materials have about the same yield strength in compression as in tension.

Let us now select a general plane-stress situation composed of the stresses σ_1 and σ_2, which may have any magnitude and be either positive or negative. Thus, $\sigma_3 = 0$ because it is a plane stress state. In order to compare the failure theories with each other, we create a rectangular reference system (Fig. 11-5) in which σ_1 is plotted on the horizontal axis and σ_2 on the vertical axis. The positive (tensile) directions are taken upward on the σ_2 axis, and to the right on the σ_1 axis.

The tension test gives the tensile yield strength S_{yt} and the compression test provides S_{yc}. As shown, these points can be located on each axis. Now, by specifying that $\sigma_2 = a\sigma_1$ and letting a vary from 0 to 1, we can calculate sets of values for each failure theory and plot the results in the first and third quadrants. A similar procedure using $\sigma_2 = -b\sigma_1$, with b varying from 0 to 1, will provide

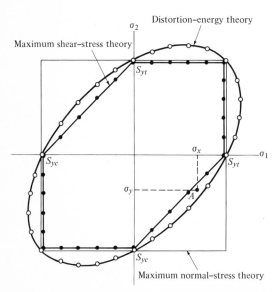

FIGURE 11-5. Comparison of three static-failure theories for ductile materials and biaxial stresses.

the necessary results for completing the paths in the second and fourth quadrants. Figure 11-5 shows the results from this procedure for the three theories of failure studied in the preceding section.

To use such a diagram, consider the two stress components σ_x and σ_y, which have been plotted to obtain point A on the diagram. Point A falls inside the limits defined by the maximum-normal-stress theory and the distortion-energy theory. So, by these two theories, the stress state $\sigma_x\sigma_y$ is considered safe. But, A falls outside the limit defined by the maximum-shear-stress theory, and so this theory predicts failure. By plotting the principal stresses in Example 11-1 on this or a similar diagram, it is easy to see the relations between the various factors of safety that were obtained.

Figure 11-6 is a chart of the same theories, in which only the first and fourth quadrants have been used. This is permissible because σ_1 and σ_2 can always be interchanged. A number of actual test points are shown on the chart with the purpose of discovering which theory is the best one to use.

An examination of the first quadrant, where σ_1 and σ_2 have the same sign, shows all points to be outside the failure lines for the maximum-normal-stress theory and the maximum-shear-stress theory. Thus, both of these theories pre-

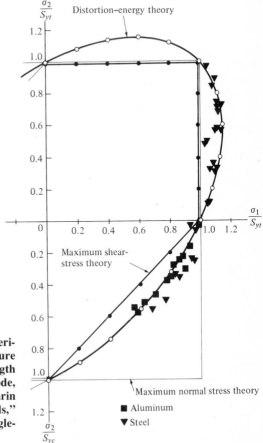

FIGURE 11-6. Chart showing how experimental results compare with three failure theories. The axes are the stress-strength ratios. Tests by Taylor, Quinney, Lode, Marin, and Stanley, as reported by Marin in Joseph Marin, "Engineering Materials," pp. 156, 157, Prentice-Hall, Inc., Englewood Cliffs, N. J., 1952.

dict smaller values for the strength, and so they are on the conservative side and are safe to use. Of course, both of these theories give the same results in this quadrant. The first quadrant also shows that the distortion-energy theory comes closest to predicting actual failure.

Examination of the data in the fourth quadrant, where σ_1 and σ_2 have opposite signs, again shows that the distortion-energy theory comes closest to predicting actual failure. Note, too, that all data points fall inside the maximum-normal-stress trajectory. This means that the maximum-normal-stress theory is a dangerous one to use when the principal stresses have opposite signs, because it will predict safety when, in fact, none exists. The fourth quadrant plot also shows that the maximum shear stress theory is always on the safe side. Of course, in evaluating any set of experimental evidence, the statistical nature of the data should be considered because there will usually be some data points that vary considerably from the normal.

In summary, we would probably use the distortion-energy theory if we wanted to find out why a part had failed in service. We would also use it in very meticulous design where the loads and stresses are in very careful control and where only a small margin of safety can be permitted. For all other analysis and design situations involving static loads and ductile materials, the maximum shear stress theory is the safest to use.

A stress situation that arises quite frequently is one which involves bending and torsion. In analyzing the safety of such a stress state, it is necessary to draw a Mohr's circle diagram to find σ_1 and σ_3, and then to use Eq. (11-5) to find the von Mises stress when the distortion-energy theory is employed. A short cut, which eliminates the use of Mohr's circle, is available for the $\sigma_x \tau_{xy}$ stress state. For this state, the von Mises stress is

$$\sigma' = \sqrt{\sigma_x^2 + 3\tau_{xy}^2} \tag{11-6}$$

Now, prove that Eq. (11-6) is true.

11-6. FAILURE THEORIES FOR BRITTLE MATERIALS   In selecting a failure theory for use with brittle materials we first observe that most brittle materials have the following characteristics:

1   The stress-strain diagram is a smooth continuous line to the failure point; failure occurs by fracture and these materials do not have a yield point.
2   The compressive strength is usually many times greater than the tensile strength.
3   The modulus of rupture S_{su}, sometimes called the ultimate shear strength, is approximately the same as the tensile strength.

The *Coulomb-Mohr theory,* also called the *internal-friction theory,* is based upon the results of two tests, the standard tension and compression tests. These experiments give the tensile and compressive strengths S_{ut} and S_{uc}. The theory is based upon plotting two Mohr's circles on the same diagram, one for each test. These are shown in Fig. 11-7. Now draw line *AB* tangent to both

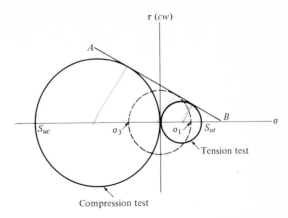

FIGURE 11-7.

circles. The Coulomb-Mohr theory states that failure occurs for any stress state that produces a circle tangent to the envelope formed by the two test circles and the tangent line. As an example, suppose a particular stress state in which the principal stresses are arranged such that $\sigma_1 > \sigma_2 > \sigma_3$. Then, the largest circle will be formed by σ_1 and σ_3. If this circle is tangent to the envelope, as shown by the dashed lines in Fig. 11-7, then failure will occur. The stresses σ_1 and σ_3 and the two strengths can be related by the equation

$$\frac{\sigma_1}{S_{ut}} + \frac{\sigma_3}{S_{uc}} = 1 \tag{11-7}$$

where both σ_3 and S_{uc} are negative quantities. For a safe stress state, a factor of safety n will exist, and Eq. (11-7) becomes

$$\frac{\sigma_1}{S_{ut}} + \frac{\sigma_3}{S_{uc}} = \frac{1}{n} \tag{11-8}$$

Figure 11-8 is a plot of the maximum-normal-stress theory and the Coulomb-Mohr theory (with test results for comparison purposes). Note that σ_3 is assumed to be zero, that σ_1 is always tension, and that σ_2 may be either a tensile or a compressive stress. It is seen that the maximum-normal-stress theory is satisfactory when both stresses are tension. Then, with one compressive stress, the Coulomb-Mohr theory is seen to be on the safe side of test results.

For pure torsion, $\sigma_2 = -\sigma_1$. This is shown on Fig. 11-8 as a 45° line, and results in two predictions for the modulus of rupture S_{su}. Based on the maximum-normal-stress theory, we see that $S_{su} = S_{ut}$. And based on the Coulomb-Mohr theory, Eq. (11-7) can be solved to give the modulus of rupture as

$$S_{su} = \frac{S_{ut} S_{uc}}{S_{ut} + S_{uc}} \tag{11-9}$$

This equation gives a very conservative estimate of S_{su} when it is solved for various cast irons in Table A-17 and the results are compared.

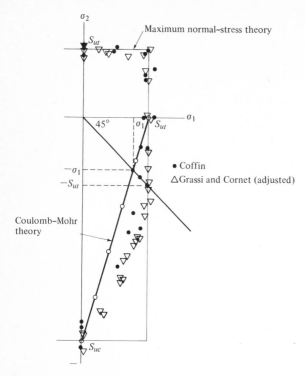

FIGURE 11-8. A plot of experimental data points from tests of gray cast iron subjected to biaxial stresses. The data were adjusted to correspond to $S_{ut} = 32$ kpsi and $S_{uc} = 105$ kpsi. Superimposed on the plot are the graphs of the maximum-normal-stress theory and the Coulomb-Mohr theory.

11-7. STRESS CONCENTRATION   In the development of the basic stress equations for tension, compression, bending, and torsion, it is assumed that no irregularities exist in the member under consideration. But, almost any structural or machine member will have one or more changes in cross section for the simple reason that other parts must be connected to them. Beams must be riveted, welded, or bolted to columns. Shafts must have keyways so that gears and pulleys may be secured to them, and shoulders so that assembled parts can be accurately located in the axial direction. Bolts have a change in cross section at the head and also where the threads are cut. Other parts may require holes, notches, and grooves, as well as other kinds of irregularities. Such discontinuities disturb the stress distribution resulting in a very sudden increase in stress in the immediate vicinity of the section change. For this reason, these discontinuities are called *stress raisers,* and the regions in which they occur are called areas of *stress concentration.*

The determination of the shape of the stress distribution near a discontinuity is such a difficult problem that only a few geometric configurations have

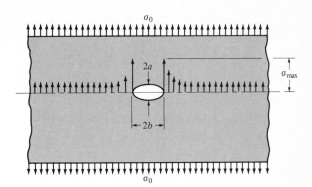

FIGURE 11-9. Stress distribution near an elliptical hole in an infinite plate loaded in tension.

actually been solved. For this reason, most stress analysts prefer to use experimental means to obtain maximum stresses.

One of the few problems which have been solved is that of an infinitely wide plate containing a small elliptical hole loaded uniformly in tension. Figure 11-9 shows the approximate stress distribution. Notice how the stress rises very abruptly near the hole. The maximum tensile stress is given by the equation

$$\sigma_{max} = \left(1 + \frac{2b}{a}\right)\sigma_0 \tag{11-10}$$

where a and b are defined in the figure. The stress σ_0 is called the *nominal stress*. Generally, the nominal stress is the stress obtained by using the basic stress equations with the net area or net cross section. In this case, the nominal stress could be written

$$\sigma_0 = \frac{F}{A} \tag{a}$$

where A is the cross-sectional area of the plate obtained by multiplying the width by the thickness and subtracting the area removed by the hole. In practice, Eq. (11-10) is usually written as

$$\sigma_{max} = K_t\sigma_0 \tag{11-11}$$

where, in this case,

$$K_t = \left(1 + \frac{2b}{a}\right) \tag{b}$$

The constant K_t is called the *theoretical* or *geometric stress concentration factor*. Its value is independent of the material of the part, depending only on the geometry involved. If the hole of Fig. 11-9 is changed to a circle, then $a = b$ and Eq. (b) becomes $K_t = 3$. Note, however, that this is for an infinite plate. For finite plates or bars, use Fig. 11-10. Figures 11-11 to 11-14 can be used to find stress-concentration factors for other geometries.

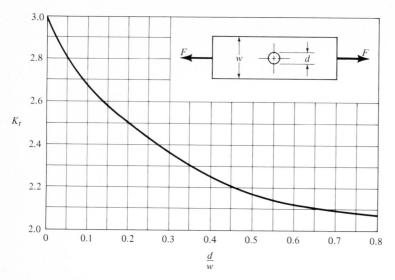

FIGURE 11-10. Stress concentration factors for a bar in tension or compression with a transverse hole. The nominal stress is $\sigma_0 = F/A$, where $A = (w - d)t$ and t is the thickness.[1]

[1] These and the factors for Figs. 11-11 to 11-14 are from R. E. Peterson, Design Factors for Stress Concentration, *Machine Design*, vol. 23, 1951 and are reproduced with the permission of the author and publisher. See also, R. E. Peterson, "Stress Concentration Factors," John Wiley & Sons, Inc., New York, 1974.

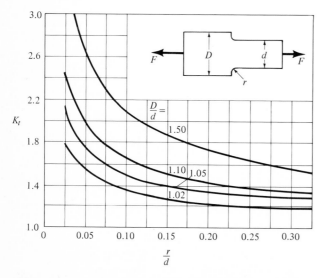

FIGURE 11-11. Stress concentration factors for a rectangular bar with fillets loaded in tension or simple compression. $\sigma_0 = F/A$, where $A = td$ and t is the thickness.

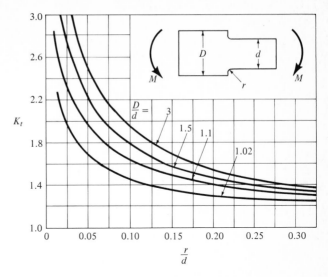

FIGURE 11-12. Stress concentration factors for a rectangular bar with fillets, loaded in bending. $\sigma_0 = Mc/I$, where $c = d/2$, $I = td^3/12$, and t is the thickness.

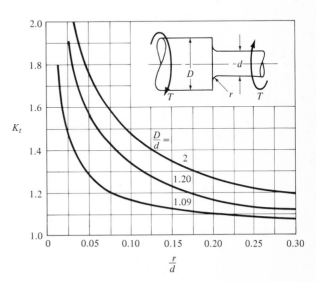

FIGURE 11-13. Stress concentration factors for a round shaft with a shoulder fillet, in torsion. $\tau_0 = Tc/J$, where $c = d/2$ and $J = \pi d^4/32$.

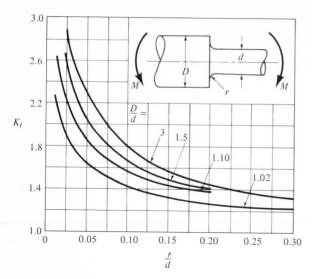

FIGURE 11-14. Stress concentration factors for a round shaft with shoulder fillet, in bending. $\sigma_0 = Mc/I$, where $c = d/2$ and $I = \pi d^4/64$.

Stress concentration factors are not usually applied to the basic stress equations if the loads are static and ductile materials are used. The reason for this is that the first load applied to the member causes local yielding at the discontinuity. This yielding redistributes the stresses and enables the gross section to carry the load.

Stress concentration factors should definitely be used when computing maximum stresses in members made of brittle materials, because these cannot yield. Usually, a reduced value of K_t is used, however, because many of the brittle materials are not particularly sensitive to notches.

11-8. FATIGUE FAILURE   In obtaining the properties of materials relating to the stress-strain diagram, the load is applied gradually, thus giving sufficient time for the strain to develop. Usually, the specimen is tested to destruction and so the stresses are applied only once. This is a static loading situation, and such conditions are closely approximated in many structural and machine members.

The condition frequently arises, however, in which the stresses vary or fluctuate between several sets of values. For example, a point on the surface of a rotating shaft subjected to the action of bending loads undergoes both tension and compression for each revolution of the shaft. A point on the shaft of an 1800-rpm electric motor loaded in bending would have a bending stress changing from tension to compression 1800 times a minute. Members are often found which have failed under the action of such stresses, and yet the most careful examination and analysis reveals that the actual maximum stresses were below the ultimate strength of the material, and frequently even below the yield strength. The most distinguishing characteristic of these failures has been that the stresses have been repeated a very large number of times. Hence, the failure is called a *fatigue* failure. And the stresses which cause such a failure are often called *fatigue stresses*.

A fatigue failure begins with a small crack. The initial crack is so minute that you can't see it, and it is even hard to find in an x-ray inspection. Usually, the crack starts at some discontinuity, such as a change in the cross section or a keyway or hole. Less obvious points at which a fatigue failure might start are inspection or stamp marks, internal cracks, or even irregularities caused by machining. Once a crack has developed, the stress-concentration effect becomes greater and the crack progresses more rapidly. As the stressed area decreases in size, the stress increases in magnitude until, finally, the remaining area fails suddenly. A fatigue failure, therefore, is characterized by two distinct areas of failure. The first of these is due to the progressive development of the crack, while the second is due to the sudden fracture. The zone of the sudden fracture is very similar in appearance to the fracture of a brittle material, such as cast iron, which has failed in tension.

When parts fail statically, they often develop a very large deflection because the stress has exceeded the yield strength. Such large deflections are quite visible, and so the part will usually be replaced before fracture actually occurs. Thus, many static failures are visible ones and give warning in advance. But a fatigue failure gives no warning; it is sudden and total and, hence, dangerous. Fatigue is a complicated phenomenon and is only partially understood. Consequently, almost all of our knowledge concerning fatigue has been acquired by experiment.

To determine the strength of materials under the action of fatigue loads, specimens are subjected to repeated or varying forces of specified magnitudes while the cycles or number of stress reversals are counted to destruction. One of the most widely used fatigue testers is the high-speed rotating-beam machine. The typical specimen used in this machine is shown in Fig. 11-15a. Before mounting, the specimen is very carefully machined and polished with a final polishing in an axial direction to eliminate all circumferential scratches. The specimen is then loaded so as to create a constant bending moment in the region in which failure is expected (Figs. 11-15b and c).

Fatigue failure is a statistical phenomenon. This makes it necessary to test a large number of specimens in order to establish the fatigue strength. The first test is usually made at a stress slightly less than the ultimate strength of the material used. At this stress level, the rotating-beam machine may only run for a few thousand revolutions before failure occurs. This number of revolutions is recorded as the number of stress reversals N along with the stress. The second specimen is tested at a stress somewhat less than the first. This process is continued and the results are plotted on an S-N diagram (Fig. 11-16). Eventually, a stress will be reached, if the specimens are of ferrous metals or alloys, which results in no failure no matter how long the machine is run. This stress level is called the *endurance limit* or *fatigue limit* and is designated as S_e. Plotting the S-N diagram on semilog or log-log paper emphasizes the break or the knee in the diagram corresponding to S_e.

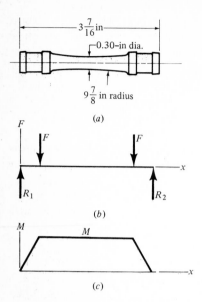

(a)

(b)

(c)

FIGURE 11-15. (a) Test specimen for the rotating-beam fatigue test; (b) loading diagram; (c) bending-moment diagram.

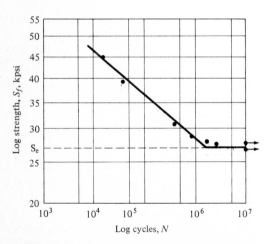

FIGURE 11-16. An S-N diagram for annealed 1040 steel. The data observed and plotted are 45 kpsi at 15 100 cycles, 39.1 kpsi at 41 300 cycles, 30.9 kpsi at 408 300 cycles, 28.9 kpsi at 897 400 cycles, 28.05 kpsi at 1 696 000 cycles, 27.4 kpsi at 2 692 000 cycles, 27.25 kpsi at 10 012 000 cycles, and 27.0 kpsi at 10 089 000 cycles. The last two specimens did not fail. The endurance limit at the knee of the graph is $S_e = 27.1$ kpsi.

The ordinate of the *S-N* diagram is called the *fatigue strength S_f*. Any statement of this strength must always be accompanied by a statement of the number of cycles N to which it corresponds. Nonferrous alloys do not have an endurance limit, and so the fatigue strength and the corresponding life in cycles is specified for them. The fatigue strength of the aluminum alloys is usually stated for a life $N = 50(10)^7$ cycles of stress reversal (see Table A-16).

11-9. ENDURANCE LIMITS The endurance limit of a structural or machine member subjected to fatigue loading is often substantially less than the endurance limit of a rotating-beam specimen made of the same material. In fact, in some cases, the reduction may be as much as 80 percent. Quantitative methods are available to help in estimating the amount of this reduction[1], but here we shall only discuss the various factors which contribute to this reduction without assigning them specific values.

The endurance limit of a member varies considerably with the quality of the surface finish of the part. In fact, the rotating-beam specimen is highly polished with the final polishing in the axial direction in an effort to get the highest possible endurance limit. Most fatigue-loaded members do not have such a fine finish because it is simply too expensive to produce them. This means that various surface irregularities may exist from which a fatigue failure can originate. The amount of reduction in endurance limit which is caused by surface finish depends upon the tensile strength of the material, as well as the quality of the finish. For high-strength steels, the reduction is about 35 percent for machined and cold-drawn parts, and nearly 80 percent for steels that have a forged finish.

The standard rotating-beam test gives the endurance limit for a specimen 0.30 in in diameter. If smaller specimens are tested, they usually give a slightly higher fatigue limit. But, a reduction of about 10 to 15 percent is found when larger specimens are tested. This size characteristic of fatigue seems to be related to the stress gradient, because there is little change in the endurance limits for various sized specimens when they are loaded in axial fatigue.

Stress concentration factors always have to be employed in the solution of problems involving the possibility of a fatigue failure. Such failures always begin at a point of stress concentration. But, it turns out that the hardened and heat-treated steels are much more sensitive to stress concentration than are the lower-strength steels. So a reduced value of the stress-concentration factor is often used when designing parts with the low-strength materials.

Temperature, the existence of residual stresses, corrosion, and metal plating all tend to reduce endurance limits, and so these effects should be inves-

[1] The following books discuss the fatigue modification factors in detail:
(1) John O. Almen and Paul H. Black, "Residual Stresses and Fatigue in Metals," pp. 24–45, McGraw-Hill Book Company, New York, 1963
(2) Robert C. Juvinall, "Stress, Strain, and Strength," pp. 226–289, McGraw-Hill Book Company, New York, 1967.
(3) Joseph E. Shigley, "Mechanical Engineering Design," pp. 254–281, McGraw-Hill Book Company New York, 1972.

tigated carefully. On the other hand, various methods of cold-working, such as shot-peening or cold-rolling, can be used to build a residual compressive stress into the surface of the member. Residual surface tensile stresses definitely weaken a member in fatigue, but residual compressive stresses increase the strength—at least up to a point. So, when properly done, such cold-working processes can be used to increase endurance limits from about 10 to as much as 30 percent.

PROBLEMS

11-1 A ductile steel specimen has a yield strength in tension and simple compression of 62 kpsi. Find factors of safety based on the maximum-normal-stress theory, the maximum-shear-stress theory, and the distortion-energy theory for the following stress states:
(a) $\tau_{xy} = 20$ kpsi ccw
(b) $\sigma_x = 15$ kpsi, $\tau_{xy} = 12$ kpsi cw
(c) $\sigma_x = 12$ kpsi, $\sigma_y = -6$ kpsi, $\tau_{xy} = 8$ kpsi cw
(d) $\sigma_x = -20$ kpsi, $\sigma_y = -20$ kpsi

11-2 The same as Prob. 11-1 for the following stress states:
(a) $\sigma_x = 20$ kpsi, $\sigma_y = -20$ kpsi
(b) $\sigma_x = 12$ kpsi, $\sigma_y = -20$ kpsi, $\tau_{xy} = 8$ kpsi ccw
(c) $\sigma_x = 18$ kpsi, $\tau_{xy} = 6$ kpsi cw
(d) $\sigma_y = -12$ kpsi, $\tau_{xy} = 12$ kpsi cw

11-3 The figure shows a $1\frac{1}{2}$-in diameter steel shaft loaded by a bending moment in the xy plane $M = 4000$ lb · in, an axial tension $F = 5000$ lb, and a torque $T = 8000$ lb · in.
(a) Determine which point, A on top of the shaft, B on the side, or C on the bottom, is critical with respect to possible failure.
(b) What should be the minimum yield strength of the shaft material in order to prevent a static failure?

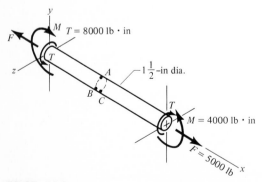

PROB. 11-3

11-4 A thin-wall pressure vessel is made of 3003-H14 aluminum alloy tubing. The vessel has an outside diameter of $2\frac{1}{2}$ in and a wall thickness of 0.065 in. What internal pressure will cause the tube walls to yield?

11-5 A tin can is made of No. 30 gauge (0.0120 in) steel and is 4 in in diameter. If the material has a yield strength of 42 kpsi, what internal pressure would cause yielding?

11-6 The figure shows a steel shaft $\frac{3}{4}$ in in diameter. Bearings at O and C have been replaced by the reactions R_1 and R_2. The shaft is loaded by the bending forces $F_A = 160$ lb and $F_B = 75$ lb, and by the torques $T_A = T_B = 900$ lb · in acting as shown. What should be the yield strength of the shaft material if a factor of safety of 2.4 is to be employed? Use the maximum-shear-stress theory.

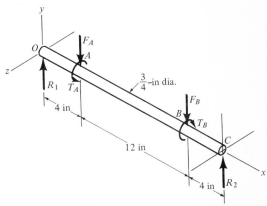

PROB. 11-6

11-7 The figure is a free-body diagram of a steel shaft supported on bearings at O and B that result in bearing reactions R_1 and R_2. The shaft is loaded by the forces $F_A = 1000$ lb and $F_C = 400$ lb, and by the equal and opposite torques $T_A = -T_C = 5400$ lb · in. The material is cold drawn AISI 1018 steel and the shaft is $1\frac{1}{4}$ in in diameter. What is the factor of safety guarding against a static failure? Use the distortion-energy theory.

PROB. 11-7

answers to selected problems

CHAPTER 3

3-5 (a) $F_O = 173$ lb; (b) $F_O = 160$ lb;
(c) $F_B = 305$ lb; (d) $F_A = 300$ lb.

3-7 (a) $F_C = 109$ lb; (b) $F_O = 18.8$ lb;
(c) $F_A = 106.7$ lb; (d) $F_O = 37$ lb.

3-9 (a) $R_O = 60$ lb; (b) $R_O = 233$ lb;
(c) $R_A = 348$ lb; (d) $R_A = 333$ lb;
(e) $R_A = 333$ lb; (f) $R_A = 333$ lb.

3-11 (a) $R_O = 88.6$ lb;
(b) $R_O = 3340$ lb;
(c) $R_O = 780$ lb;
(d) $R_O = 300$ lb;
(e) $R_O = 360$ lb;
(f) $R_A = 291$ lb.

3-13 (a) 7270 psi, $8.72(10)^{-3}$ in;
(b) 8190 psi, 0.0197 in;
(c) 2330 psi, $8.14(10)^{-3}$ in;
(d) 2560 psi, $5.98(10)^{-3}$ in.

3-15 (a) 392 MPa, 4.55 nm,
1.90 nm/m, -0.554 nm/m;
(b) 172 MPa, 5.82 mm,
2.43 mm/m, -0.810 mm/m;
(c) 142 MPa, 1.79 mm,
0.746 mm/m, -0.227 mm/m.

3-17 0.178 in, 0.242 in, 0.155 in,
0.160 in.

3-19 (a) $\frac{9}{64} \times \frac{9}{16}$ in; (b) $\frac{3}{8} \times 1\frac{1}{2}$ in;
(c) $\frac{9}{32} \times 1\frac{1}{8}$ in; (d) $\frac{11}{64} \times \frac{11}{16}$ in.

3-21 2.91 in.

3-23 (a) 3750 lb, 0.600; (b) 5000 lb,
0.500; (c) 3750 lb, 0.429;
(d) 3750 lb, 0.429.

3-25 (a) 7290 lb, 0.576; (b) 7290 lb,
0.345; (c) 8440 lb, 0.500;
(d) $14.6(10)^3$ lb, 0.611.

3-27 (a) 4290 lb, 0.550; (b) 4290 lb,
0.550; (c) 4290 lb, 0.550;
(d) 6250 lb, 0.471.

3-29 (a) $13.8(10)^3$ lb; (b) 6240 lb;
(c) $15.5(10)^3$ lb; (d) $14.2(10)^3$ lb.

3-31 $F = 5100$ lb.

3-33 $F = 32.7(10)^3$ lb.

3-35 Shear of bolt 3, bearing on bolt
5.52, bearing on members 3.90,
and tension in members 8.45.

3-37 $n = 2.84$.

3-39 9600 lb, 9600 lb, 6500 lb, 19.2(10)³ lb.

3-41 7210 lb, 33.7(10)³ lb, 18.0(10)³ lb, 9600 lb.

3-43 45.6 kpsi.

3-45 1340 psi, 2020 psi, 5220 psi.

3-47 Shear of bolt, 3.54; bearing on bolt, 6.25; bearing on member, 4.09; tension in member, 8.16.

CHAPTER 4

4-1 (a) 120 lb, 120 lb; (b) 62 lb, 138 lb; (c) 144 lb, 36 lb; (d) 157 lb, 23.3 lb.

4-3 (a) 400 lb, 600 lb; (b) 1200 lb, 1800 lb; (c) 540 lb, 60 lb; (d) 170 lb, 1370 lb.

4-5 (a) 248 lb, 232 lb; (b) 120 lb, 280 lb; (c) 180 lb, 220 lb; (d) 117 lb, 118 lb.

4-7 (a) 830 kN, 1070 kN; (b) 2295 kN, 855 kN; (c) 150 kN, 2850 kN; (d) 1725 kN, 3475 kN.

4-9 (a) 1.8, 3.2, 4.2, 4.8; (b) 1.8, 1, 0.2, −0.2; (c) 1.4, 1.0, 0.6; (d) 2, 1.2, −0.4

4-11 (a) −4, 10, −10, −8; (b) −4, 0, −8; (c) −7, −5, −1; (d) −7, −5, −3, 3.

4-13 $y' = -2/x^2$, $y' = 2bx$, $y' = 1 - 6x$, $y' = -3/x^4$, $y' = 2\pi x$.

4-15 (a) 200 lb, 2700 lb · in; (b) 250 lb, 3800 lb · in; (c) 200 lb, 3280 lb · in; (d) $F_1 + F_2$, $aF_1 + (a + b)F_2$.

4-17 (a) 360 lb, 1620 lb · ft; (b) 1300 lb, 12 100 lb · ft; (c) 960 lb, 9360 lb · ft; (d) 1160 lb, 10 680 lb · ft.

4-19 (a) $M = -480$ lb · in at $x = 4$ in; (b) $M = -10$ lb · in at $x = 2.5$ in; (c) $M_{max} = 350$ lb · ft; (d) $M_{max} = 420$ lb · ft.

4-21 (a) $M_{max} = 2560$ lb $\cdot$ ft;
 (b) $M = -480$ lb $\cdot$ ft at $x = 4$ ft;
 (c) $M_{max} = 1920$ lb $\cdot$ ft;
 (d) $M = -240$ lb $\cdot$ in at $x = 30$ in.

4-23 (a) $M = 512$ lb $\cdot$ in at $x = 10$ in;
 (b) $M = 150$ lb $\cdot$ in at
 $x = 12.5$ in;
 (c) $M_{max} = 260$ lb $\cdot$ in;
 (d) $M = -600$ lb $\cdot$ in at $x = 10$ in.

4-25 $V(O \text{ to } A) = -(w/2)(2x + l)$,
 $V(A \text{ to } B) = w(l - x)$, $M(O \text{ to } A) = -(wx/2)(x + l)$, $M(A \text{ to } B) = (w/2)(2lx - x^2 - l^2)$,
 $M_{max} = -wl^2/6$ at A.

4-27 $V(O \text{ to } A) = -wx$, $V(A \text{ to } B) = (w/2)(l - 2x)$, $V(B \text{ to } C) = w(l - x)$, $M(O \text{ to } A) = -wx^2/2$, $M(A \text{ to } B) = (w/6)(3lx - 3x^2 - l^2)$,
 $M = -wl^2/18$ at A, $M = -wl^2/24$ at $x = l/2$.

4-29 $M_{max} = 325$ lb $\cdot$ ft at $x = 7.14$ ft;
 $M_A = -360$ lb $\cdot$ ft.

4-31 $M_A = M_B = -500$ lb $\cdot$ in, $M = 0$
 at $x = 20$ in.

4-33 $R_1 = 2400$ N, $M_1 = -7200$ N $\cdot$ m,
 $V(O) = 2400$ N, $V(l) = 0$,
 $M(O) = -7200$ N $\cdot$ m, $M(l) = 0$.

4-35 $M_{max} = 90$ N $\cdot$ m at $x = 300$ mm.

CHAPTER 5

5-1 (a) $c_1 = 1.389$ in, $c_2 = 1.111$ in,
 $I = 3.76$ in⁴, $\sigma_A = 2950$ psi,
 $\sigma_B = -3690$ psi;
 (b) $c_1 = 0.594$ in, $c_2 = 0.906$ in,
 $I = 0.293$ in⁴, $\sigma_A = 30.9$ kpsi,
 $\sigma_B = 22.4$ kpsi, $\sigma_C = -11.7$ kpsi,
 $\sigma_D = -20.3$ kpsi.

5-3 (a) $c_1 = 1.708$ in, $c_2 = 2.292$ in,
 $I = 10.99$ in⁴, $\sigma_A = 2.08$ kpsi,
 $\sigma_B = 1.63$ kpsi, $\sigma_C = -1.10$ kpsi,
 $\sigma_D = -1.55$ kpsi;
 (b) $c_1 = 1.354$ in, $c_2 = 2.646$ in,
 $I = 7.97$ in⁴, $\sigma_A = 3.32$ kpsi,
 $\sigma_B = -1.03$ kpsi,
 $\sigma_C = -1.70$ kpsi.

5-5 (a) $c_1 = 0.466$ in, $c_2 = 1.034$ in,
$I = 0.139$ in^4, $\sigma_A = 74.4$ kpsi,
$\sigma_B = -15.5$ kpsi,
$\sigma_C = -33.5$ kpsi;
(b) $c_1 = c_2 = 0.75$ in,
$I = 0.289$ in^4, $\sigma_A = 26.0$ kpsi,
$\sigma_B = 17.3$ kpsi, $\sigma_C = -17.3$ kpsi,
$\sigma_D = -26.0$ kpsi.

5-7 (a) $c_1 = 4.417$ in, $c_2 = 8.07$ in,
$I = 353.9$ in^4; (b) $c_1 = c_2 = 6$ in,
$I = 212.2$ in^4.

5-9 (a) $c_1 = 2.18$ in, $c_2 = 11.82$ in,
$I = 288.9$ in^4; (b) Use the
nominal flange width;
$c_1 = c_2 = 6.5$ in, $I = 756$ in^4.

5-11 (a) $x = 30$ in, $y = -533$ in;
(b) $x = 20$ in, $y = 2130$ in.

5-13 902 lb/ft, 7660 lb/ft, 5690 lb/ft,
3150 lb/ft.

5-15 Shear stresses are all maximum
on neutral axis.
(a) $\sigma_{max} = 14\ 200$ psi on bottom
at A, $\tau_{max} = 890$ psi from A to B;
(b) $\sigma_{max} = 12\ 000$ psi at A on top,
$\tau_{max} = 750$ psi from O to B;
(c) $\sigma_{max} = 3750$ psi on bottom at
midbeam, $\tau_{max} = 900$ psi at A and B;
(d) $\sigma_{max} = 2700$ psi on top at A,
$\tau_{max} = 562$ psi at A.

5-17 Nail spacing is 1.54 in. This
might cause splitting and hence is
not recommended.

5-19 Use 11-in spacing.

5-21 (a) 94 psi; (b) 110 psi.

5-23 (a) $\tau = 22\ 700$ psi,
$\sigma = 12\ 000$ psi;
(b) $\tau = 14\ 000$ psi,
$\sigma = 14\ 800$ psi.

5-25 (a) $c_1 = 55.4$ mm, $c_2 = 94.6$ mm,
$I = 1.858(10)^{-5}$ m^4,
$\sigma_A = 2.98$ MPa,
$\sigma_B = -5.09$ MPa;
(b) $c_1 = 43$ mm, $c_2 = 117$ mm,
$I = 2.93(10)^{-5}$ m^4,
$\sigma_A = 1.47$ MPa,
$\sigma_B = -3.99$ MPa.

CHAPTER 6

6-1 (a) 1.00 in; (b) 4.65 in.

6-3 (a) 48.2; (b) 69.5.

6-5 (a) 47 kpsi; (b) 20.3 lb.

6-7 18.6 kpsi, 15.9 kpsi.

6-9 0.83 percent.

6-11 $\tau_{max} = 10.1$ kpsi, $\theta_{AB} = 0.674°$, $\theta_{BD} = 4.72°$.

6-13 $\sigma = +8.66$ kpsi, $\tau = 5$ kpsi ccw; $\sigma = 10$ kpsi, $\tau = 0$; $\sigma = 5$ kpsi, $\tau = 8.66$ kpsi cw.

6-15 (a) 11.4 lb/in; (b) $\tau = 113$ kpsi.

6-17 (a) 0.855 lb; (b) 9.23 lb; (c) 2.1 lb/in; (d) 6.75 in.

6-19 8310 psi.

6-21 10.9 kpsi.

6-23 Use $\frac{1}{4}$-in welds.

6-25 0.474 rad, 130 MPa.

CHAPTER 7

7-11 (a) -0.451 in; (b) -1.41 in.

7-13 (a) -0.257 in; (b) -0.167 in.

7-15 $b = 0.186$ in, $h = 1.49$ in; 32.6 kpsi.

7-17 2.16 in; when w and I, in terms of b and h, are substituted in the deflection formula, b cancels. So the thickness doesn't enter into the problem.

7-19 $y_C = -0.156$ in.

7-21 $y_{max} = 0.0148$ in.

7-23 100×500 mm, 9.69 MPa.

CHAPTER 8

8-1 $R_O = W/3$, $R_B = 2W/3$.

8-3 Tube, $\sigma = -4.4$ kpsi; bar, $\sigma = -12.9$ kpsi.

8-5 32 in from the left end.

8-7 515 lb.

8-9 $7\frac{1}{3}$ ft.

8-11 $T_O = 0.6T$, $T_B = 0.4T$.

CHAPTER 9

9-1 $P_{cr} = 4\pi^2 EI/l^2$.

9-3 2.

9-5 (a) 220 lb; (b) 1160 lb; (c) 231 lb; (d) 1600 lb.

9-7 $\dfrac{P_{cr}}{A} = 54(10)^3 - 2.46\left(\dfrac{l}{k}\right)^2.$

9-9 (a) $\tfrac{3}{8}$ in; (b) $\tfrac{1}{2}$ in; (c) $\tfrac{7}{8}$ in.

9-11 4.62.

9-13 (a) 197 000 lb; (b) 96 000 lb.

9-15 X-X direction, 75 000 lb; Y-Y direction, 183 000 lb.

9-17 889 000 lb.

CHAPTER 10

10-1 -1390 psi, $+1180$ psi.

10-3 $+16.7$ kpsi, -12.3 kpsi.

10-5 $+175$ psi, -223 psi.

10-7 At P, ± 2460 psi; at Q, 2500 psi and -1320 psi.

10-9 -24.9 kpsi, $+21.6$ kpsi.

10-11 -31.6 kpsi, $+24.2$ kpsi.

10-13 26 kpsi, -24.8 kpsi.

10-15 $\sigma_u = -100$ psi, $\sigma_v = -100$ psi, $\tau_{uv} = 200$ psi cw.

10-17 $\sigma_u = -273$ psi, $\sigma_v = -127$ psi, $\tau_{uv} = 273$ psi cw.

10-19 $\sigma_1 = 14.43$ kpsi at $29°$ ccw from x.

10-21 $\sigma_1 = 8.48$ kpsi at $22.5°$ cw from x.

10-23 (a) $\sigma_1 = 11.4$ kpsi, $\sigma_2 = -1.4$ kpsi, $\phi = 19.4°$ cw to σ_1; (b) $\sigma_1 = 0$ kpsi, $\sigma_2 = -10$ kpsi, $\phi = 26.6°$ ccw to σ_1 axis; (c) $\sigma_1 = 14$ kpsi, $\sigma_2 = 1$ kpsi, $\phi = 33.7°$ cw to σ_1 axis.

10-25 (a) $\sigma_1 = 3.2$ kpsi, $\sigma_2 = -21.2$ kpsi, $\phi = 55°$ ccw to σ_1 axis; (b) $\sigma_1 = 43.5$ kpsi, $\sigma_2 = -7.5$ kpsi, $\phi = 45°$ ccw to σ_1 axis; (c) $\sigma_1 = 36$ kpsi, $\sigma_2 = -4$ kpsi, $\phi = 71.55°$ cw to σ_1 axis.

10-27 (a) $\sigma_1 = 32$ kpsi, $\sigma_2 = -32$ kpsi, $\phi = 45°$ ccw from σ_1 to x; (b) $\sigma_u = -27.7$ kpsi, $\tau_{uv} = 16$ kpsi, ccw, $\sigma_v = 27.7$ kpsi, $\tau_{vu} = 16$ kpsi cw, $2\phi = 120°$ ccw from x to u.

10-29 $\sigma_u = -4.4$ kpsi, $\sigma_v = 16.4$ kpsi,
$\tau_{uv} = 6$ kpsi cw.

10-31 **(a)** $\sigma_x = -3340$ psi,
$\tau_{xy} = 1690$ psi ccw;
(b) $\sigma_1 = 710$ psi,
$\sigma_2 = -4050$ psi, $\phi = 67.35°$ from
x ccw to σ_1.

10-33 **(a)** At A, $\sigma_x = 10.6$ kpsi,
$\tau_{xz} = 7.99$ kpsi ccw; at B,
$\tau_{xy} = 7.60$ kpsi ccw; **(b)** At A,
$\sigma_1 = 14.89$ kpsi,
$\sigma_2 = -4.29$ kpsi, $\phi = 28.2°$ ccw
from x to σ_1; at B,
$\sigma_1 = 7.60$ kpsi, $\sigma_2 = -7.60$ kpsi,
$\phi = 45°$ ccw from x to σ_1.

10-35 At A, $\sigma_1 = 133.8$ MPa,
$\sigma_2 = -38.3$ MPa, $\phi = 28.2°$ ccw
from x to σ_1; at B,
$\sigma_1 = -\sigma_2 = 68.4$ MPa,
$\phi = 45°$ ccw from x to σ_1.

CHAPTER 11
11-1 **(a)** 3.1, 1.55, 1.79; **(b)** 2.86,
2.19, 2.42; **(c)** 4.13, 2.58, 2.95;
(d) 3.1, 3.1, 1.79.
11-3 Point C; $S_y(\min) = 25.7$ kpsi.
11-5 292 psi.
11-7 $n = 1.68$.

NOTE ADDED IN PROOF   This note provides an opportunity to append two additional tables that I had not worked out during the writing of the original manuscript. The two tables shown below greatly simplify the computation of stresses in SI units because they automatically provide the prefix of the result. Similar tables are easy to work out for beam deflections.

Preferred SI units for bending stress $\sigma = Mc/I$ and torsion stress $\tau = Tr/J$.

M, T	I, J	c, r	σ, τ
N·m	m⁴	m	Pa
N·m	cm⁴	cm	MPa
N·m	mm⁴	mm	GPa
kN·m	m⁴	m	kPa
kN·m	cm⁴	cm	GPa
kN·m	mm⁴	mm	TPa
N·mm	mm⁴	mm	MPa
kN·mm	mm⁴	mm	GPa

Preferred SI units for axial stress $\sigma = F/A$ and direct shear stress $\tau = F/A$.

F	A	σ, τ
N	m²	Pa
N	mm²	MPa
kN	m²	kPa
kN	mm²	GPa

appendices

TABLE A-1
Standard SI prefixes.[1,2]

name	symbol	factor
tera	T	$1\ 000\ 000\ 000\ 000 = 10^{12}$
giga	G	$1\ 000\ 000\ 000 = 10^{9}$
mega	M	$1\ 000\ 000 = 10^{6}$
kilo	k	$1\ 000 = 10^{3}$
hecto*	h	$100 = 10^{2}$
deka*	da	$10 = 10^{1}$
deci*	d	$0.1 = 10^{-1}$
centi*	c	$0.01 = 10^{-2}$
milli	m	$0.001 = 10^{-3}$
micro	μ	$0.000\ 001 = 10^{-6}$
nano	n	$0.000\ 000\ 001 = 10^{-9}$
pico	p	$0.000\ 000\ 000\ 001 = 10^{-12}$
femto	f	$0.000\ 000\ 000\ 000\ 001 = 10^{-15}$
atto	a	$0.000\ 000\ 000\ 000\ 000\ 001 = 10^{-18}$

* Not recommended but sometimes encountered.

[1] If possible, use multiple and submultiple prefixes in steps of 1000. For example, specify length in mm, m, or km, say. In a combination unit, use prefixes only in the numerator. For example, use meganewton per square metre (MN/m^2), but not newton per square centimetre (N/cm^2), nor newton per square millimetre (N/mm^2).

[2] Spaces are used in SI instead of commas to group numbers to avoid confusion with the practice in some European countries of using commas for decimal points.

TABLE A-2
Conversion of English units to SI units.

to convert from	to	multiply by accurate	common
foot (ft)	metre (m)	3.048 000*E − 01	0.305
foot-pound-force (ft · lbf)	joule (J)	1.355 818 E + 00	1.35
foot-pound-force/second (ft · lbf/s)	watt (W)	1.355 818 E + 00	1.35
inch (in)	metre (m)	2.540 000*E − 02	0.0254
gallon (gal US)	$metre^3$ (m^3)	3.785 412 E − 03	0.003 78
horsepower (hp)	kilowatt (kw)	7.456 999 E − 01	0.746
mile (mi US Statute)	kilometre (km)	1.609 344*E + 00	1.610
pascal (Pa)†	$newton/metre^2$ (N/m^2)	1.000 000*E + 00	1
poundal (pdl)	newton (N)	1.382 550 E − 01	0.138
pound-force (lbf avoirdupois)	newton (N)	4.448 222 E + 00	4.49
pound-mass (lbm avoirdupois)	kilogram (kg)	4.535 924 E − 01	0.454
pound-force/$foot^2$ (lbf/ft^2)	pascal (Pa)	4.788 026 E + 01	47.9
pound-force/$inch^2$ (psi)	pascal (Pa)	6.894 757 E + 03	6890
slug	kilogram (kg)	1.459 390 E + 01	14.6
ton (short 2000 lbm)	kilogram (kg)	9.071 847 E + 02	907

* Exact.
† The pascal (Pa) is the new name for the unit of pressure and stress and takes the place of the newton per square metre.

TABLE A-3
Conversion of SI units to English units.

to convert from	to	multiply by accurate	common
joule (J)	foot-pound-force (ft · lbf)	7.375 620 E − 01	0.737
kilogram (kg)	slug	6.852 178 *E* − 02	0.0685
kilogram (kg)	pound-mass (lbm avoirdupois)	2.204 622 E + 00	2.20
kilogram (kg)	ton (short 2000 lbm)	1.102 311 E − 03	0.001 10
kilometre (km)	mile (mi US Statute)	6.213 712 E − 01	0.621
kilowatt (kW)	horsepower (hp)	1.341 022 E + 00	1.34
metre (m)	foot (ft)	3.280 840 E + 00	3.28
metre (m)	inch (in)	3.937 008 E + 02	39.4
$metre^3$ (m^3)	gallon (gal US)	2.641 720 E + 02	264.
newton (N)	poundal (pdl)	7.233 011 E + 00	7.23
newton (N)	pound-force (lbf avoirdupois)	2.248 089 E − 01	0.225
pascal (Pa)	pound-force/$foot^2$ (lbf/ft^2)	2.088 543 E − 02	0.0209
pascal (Pa)	pound-force/$inch^2$ (psi)	1.450 370 E − 04	0.000 145
watt (W)	foot-pound-force/second (ft · lbf/s)	7.375 620 E − 01	0.737

TABLE A-4
The Greek alphabet.

Alpha	A	α	ɑ	Nu	N	ν		
Beta	B	β	β	Xi	Ξ	ξ		
Gamma	Γ	γ		Omicron	O	o		
Delta	Δ	δ	∂	Pi	Π	π		
Epsilon	E	ϵ	ε	Rho	P	ρ		
Zeta	Z	ζ		Sigma	Σ	σ	ς	
Eta	H	η		Tau	T	τ		
Theta	Θ	θ	ϑ	Upsilon	Y	υ		
Iota	I	ι		Phi	Φ	ϕ		φ
Kappa	K	κ	ϰ	Chi	X	χ		
Lambda	Λ	λ		Psi	Ψ	ψ		
Mu	M	μ		Omega	Ω	ω		

TABLE A-5
Differentials.

$d(ax) = a\ dx$

$d(x + y) = dx + dy$

$d(xy) = x\ dy + y\ dx$

$d\left(\dfrac{x}{y}\right) = \dfrac{y\ dx - x\ dy}{y^2}$

$d(x^2) = 2x\ dx$

$d(x^3) = 3x^2\ dx$

$d(x^n) = nx^{n-1}\ dx$

$d(e^x) = e^x\ dx$

$d(e^{ax}) = ae^{ax}\ dx$

$d(a^x) = a^x \ln a\ dx$

$d(\ln x) = \dfrac{dx}{x}$

$d(\sin x) = \cos x\ dx$

$d(\cos x) = -\sin x\ dx$

$d(\tan x) = \sec^2 x\ dx$

TABLE A-6
Integrals.

$\displaystyle\int a\ dx = a\int dx = ax$

$\displaystyle\int y\ dx = xy - \int x\ dy$

$\displaystyle\int x\ dx = \frac{x^2}{2}$

$\displaystyle\int x^2\ dx = \frac{x^3}{3}$

$\displaystyle\int x^3\ dx = \frac{x^4}{4}$

$\displaystyle\int x^n\ dx = \frac{x^{n+1}}{n+1}$

$\displaystyle\int \frac{dx}{x} = \ln x$

$\displaystyle\int e^x\ dx = e^x$

$\displaystyle\int e^{ax}\ dx = \frac{e^{ax}}{a}$

$\displaystyle\int \ln x\ dx = x \ln x - x$

$\displaystyle\int a^x \ln a\ dx = a^x$

$\displaystyle\int \sin x\ dx = -\cos x \text{ or } 1 - \cos x$

$\displaystyle\int \cos x\ dx = \sin x \text{ or } 1 - \sin x$

$\displaystyle\int \tan x\ dx = -\ln \cos x$

$\displaystyle\int \sin^2 x\ dx = \frac{x}{2} - \frac{1}{4}\sin 2x$

$\displaystyle\int \cos^2 x\ dx = \frac{x}{2} + \frac{1}{4}\sin 2x$

centroids and center of gravity

1. INTRODUCTION In Fig. 1*a*, the water in a lake of depth *h* is held back by a concrete dam. The pressure of the water on the dam is proportional to its depth and hence is a triangular distribution of force, as shown. It is often possible to simplify an engineering problem by replacing a force distribution or pressure acting on an area by a single resultant force. We then need to ask the question: Where must this force be applied in order to have exactly the same effect?

To answer the question, we show the same triangular pressure distribution in Fig. 1*b* resting on a fulcrum. When this triangle is adjusted until it is in perfect balance, then the distance $\bar{y}$ locates the line of action along which the single resultant force must be placed. This line of action is called the *centroidal axis* of the triangle.

In solving engineering problems, we frequently encounter the problem that forces are distributed in some manner over a line, an area, or a volume. The resultant of these distributed forces is usually not too difficult to determine. In order to have the same effect, this resultant must act at the *centroid* of the system. Thus, *the centroid of a system is a point at which a system of distributed forces may be considered concentrated with exactly the same effect.*

In many cases the centroid or the centroidal axes of a system can be located by symmetry. If an area or a volume has either a line or a plane of symmetry, then the centroid is on the line or plane. Therefore, in locating centroids, always take advantage of whatever symmetry happens to exist.

2. POINTS In Fig. 2, a beam 2 loaded by a concentrated force *F* is bolted to two columns 1, using three bolts at each end. In analyzing such a beam, it is necessary to know the exact locations of the centroids of the bolt groups.

Since the centroid of a single bolt is a point at its center, we can generalize the problem to that of finding the centroid of a group of points. To do this, we need to associate a weighting with each point which, in this case, would be the area of each bolt. In other cases, the weighting associated with each point in a group might be the weight or the mass associated with the point. In these cases, the centroid of the point group would be called, respectively, the *center of gravity* or the *center of mass*.

In Fig. 3, four weights, W_1, W_2, W_4, and W_5 are attached to a straight horizontal rod whose weight W_3 is shown acting at the center of gravity of the rod. We wish to find the center of gravity of the system.

Assume the center of gravity of the system is located at *G* a distance $\bar{x}$ ft from the left end. The total weight is

$$W = W_1 + W_2 + W_3 + W_4 + W_5$$

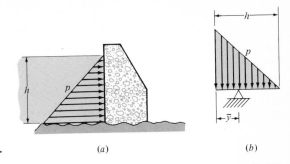

FIGURE 1.   (*a*)   (*b*)

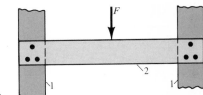

FIGURE 2.

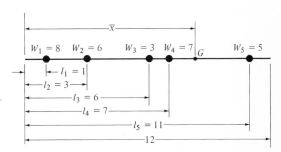

FIGURE 3. The distances *l* are in feet; the weights *W* are in pounds.

This weight when multiplied by the distance $\bar{x}$ must balance or cancel the sum of the individual weights multiplied by their respective distances from the left end. In other words,

$$W\bar{x} = W_1 l_1 + W_2 l_2 + W_3 l_3 + W_4 l_4 + W_5 l_5$$

Solving for $\bar{x}$ gives

$$\bar{x} = \frac{W_1 l_1 + W_2 l_2 + W_3 l_3 + W_4 l_4 + W_5 l_5}{W_1 + W_2 + W_3 + W_4 + W_5} \tag{1}$$

Using the given values, we get

$$\bar{x} = \frac{(8)(1) + (6)(3) + (3)(6) + (7)(7) + (5)(11)}{8 + 6 + 3 + 7 + 5} = 5.10 \text{ ft} \qquad Ans.$$

Figures 4*a* and *b* show the location of five points in a bolt group. The "weights" are the areas *A* of the bolts. The centroid *G* of the group, shown in Fig. 4*c*, requires that we find two centroidal distances. As shown, these are the distances $\bar{x}$ and $\bar{y}$ measured from the *y* and *x* axes, respectively. The procedure is the same as indicated by Eq. (1).

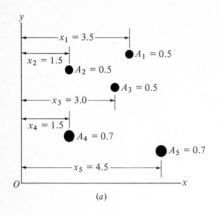
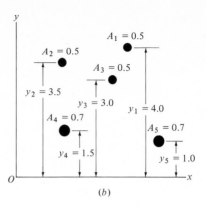

(a)

(b)

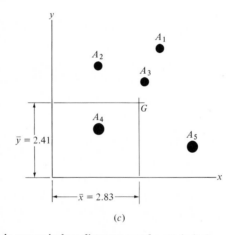

(c)

FIGURE 4. Areas A are in square inches; distances x and y are in inches.

Thus, for $\bar{x}$, we have

$$\bar{x} = \frac{A_1x_1 + A_2x_2 + A_3x_3 + A_4x_4 + A_5x_5}{A_1 + A_2 + A_3 + A_4 + A_5}$$

$$= \frac{(0.5)(3.5) + (0.5)(1.5) + (0.5)(3) + (0.7)(1.5) + (0.7)(4.5)}{0.5 + 0.5 + 0.5 + 0.7 + 0.7}$$

$$= 2.83 \text{ in} \quad Ans.$$

Next, using Fig. 4b, we find $\bar{y}$ to be

$$\bar{y} = \frac{A_1y_1 + A_2y_2 + A_3y_3 + A_4y_4 + A_5y_5}{A_1 + A_2 + A_3 + A_4 + A_5}$$

$$= \frac{(0.5)(4.0) + (0.5)(3.5) + (0.5)(3) + (0.7)(1.5) + (0.7)(1)}{0.5 + 0.5 + 0.5 + 0.7 + 0.7}$$

$$= 2.41 \text{ in} \quad Ans.$$

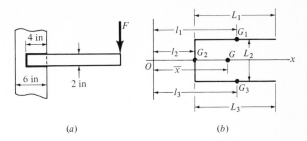

FIGURE 5. (*a*) Cantilever welded to a column using fillet welds; (*b*) geometry of the weld group.

We can now better appreciate that Eq. (1) can be expressed in the more general form

$$\bar{x} = \frac{\sum_{i=1}^{i=N} W_i l_i}{\sum_{i=1}^{i=N} W_i} \tag{2}$$

Here the Greek letter Σ (sigma) means to add up or *sum*. Thus, for example, the numerator means to sum all the $W_i l_i$'s beginning with $i = 1$ and ending with $i = N$.

3. LINES Figure 5*a* illustrates a weldment consisting of a cantilever welded to a column. The weld, called a fillet weld, has a triangular cross section. In analyzing this cantilever, it is necessary to locate the centroid of the weld group. Such a problem can be generalized to that of finding the centroid of a group of lines.

In Fig. 5*b*, the three welds of length L_1, L_2, and L_3 are shown to a larger scale. The points G_1, G_2, and G_3 are the centroids of the individual welds. The weld group has an axis of symmetry and as the figure shows, we have elected to make this axis the x axis, thus making $\bar{y} = 0$. The distances l_1, l_2, and l_3 are measured from the origin O to the individual centroids. The weighting associated with a line is simply the length of the line. Thus, Eq. (2) is written

$$\bar{x} = \frac{L_1 l_1 + L_2 l_2 + L_3 l_3}{L_1 + L_2 + L_3} \tag{3}$$

For the dimensions shown in Fig. 5*a* with the origin O at the left side of the column, we have $l_1 = l_3 = 4$ in and $l_2 = 2$ in. Therefore,

$$\bar{x} = \frac{(4)(4) + (2)(2) + (4)(4)}{4 + 2 + 4} = 3.60 \text{ in} \qquad Ans.$$

4. AREAS In designing the connections to load-carrying members in important structures and machines, engineers and designers will often go to great lengths to assure that the lines of action of the transmitted forces coincide with the centroidal axes of the sections. This assurance eliminates many unwanted difficulties.

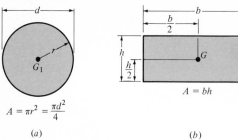

$$A = \pi r^2 = \frac{\pi d^2}{4}$$

$$A = bh$$

(a)   (b)

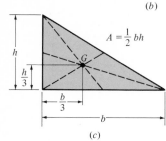

$$A = \frac{1}{2}bh$$

(c)

FIGURE 6. The centroids of the circle, rectangle, and triangle.

Many of the sections encountered are composed of combinations of circles, rectangles, and triangles. Since the centroids of these can be found by symmetry, we can use the methods already developed to obtain the centroids of various combinations of these areas. Figure 6 shows the location of the centroids of the circle, rectangle, and triangle – all obtained from symmetry. Note that the centroid of a triangle lies at the intersection of the medians.

To illustrate, consider the area in Fig. 7a as composed of a triangle, a rectangle, and a circular hole. These are delineated in Fig. 7b. The area of the triangle is

$$A_1 = \frac{1}{2}bh = \left(\frac{1}{2}\right)(6)(3) = 9 \text{ in}^2$$

The area of the rectangle is

$$A_2 = bh = (6)(5) = 40 \text{ in}^2$$

And the area of the circle is

$$A_3 = \frac{\pi d^2}{2} = \frac{\pi (2)^2}{4} = 3.14 \text{ in}^2$$

The total area is

$$A = A_1 + A_2 - A_3 = 9 + 40 - 3.14 = 45.86 \text{ in}^2$$

where A_3 is negative because it is a hole.

We next locate the three centroids G_1, G_2, and G_3 of the three areas, respectively, as shown in Fig. 7b. Then, using Eq. (2) for the x axis, we get

$$\bar{x} = \frac{A_1 x_1 + A_2 x_2 - A_3 x_3}{A}$$

$$= \frac{(9)(2) + (40)(4) - (3.14)(6)}{45.86} = 3.47 \text{ in}$$

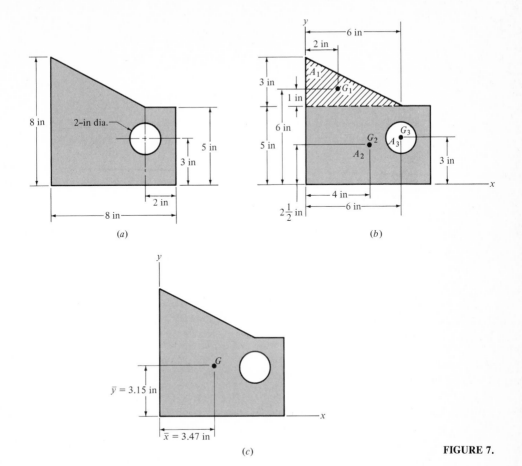

(a)

(b)

(c)

FIGURE 7.

Similarly, for y,

$$\bar{y} = \frac{A_1y_1 + A_2y_2 - A_3y_3}{A}$$

$$= \frac{(9)(6) + (40)(2\frac{1}{2}) - (3.14)(3)}{45.86} = 3.15 \text{ in}$$

The location is shown to scale in Fig. 7c.

5. SUMMATIONS Suppose we have eight separate forces W_1 through W_8 arranged on a line as shown in Fig. 8a. Mentally, we can add these forces and replace them by the single resultant froce $W = 38$, as shown in Fig. 8b. This process of adding or summing is called *summation* by mathematicians, and they would express it, using the Greek letter sigma, as

$$W = \sum_{i=1}^{i=8} W_i \tag{4}$$

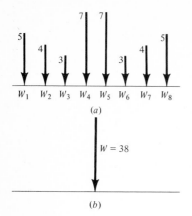

(a)

$W = 38$

(b) **FIGURE 8.**

which means to add up all the W's beginning with W_1 and ending with W_8. Expanding Eq. (4) and carrying out the summation gives

$$W = \sum_{i=1}^{i=8} W_i = W_1 + W_2 + W_3 + W_4 + W_5 + W_6 + W_7 + W_8$$

$$= 5 + 4 + 3 + 7 + 7 + 3 + 4 + 5 = 38 \quad Ans.$$

Sometimes the things that we want to sum are distributed continuously. An approximate summation can always be made of a continuous distribution by dividing the distribution into small segments. To show how this works, let us take the semicircle of Fig. 9a, which is a continuous distribution from A to B. We want to find the area.

Using the method shown in Fig. 9b, we divide the base of the semicircle into 2-in portions, each of which is to be the base of a rectangle. We are going to form four rectangles whose total area will approximate the area of the semicircle. Let us choose to make the height of each rectangle equal to the average ordinate to the semicircle. Rectangle 2, as an example, has ordinates of 3.48 in and 4 in. The average of these two is 3.74 in, and so rectangle 2 is made 3.74 in high. Since the base is 2 in wide, we find $A_2 = (2)(3.74) = 7.48$ in², as shown. Then using Eq. (4), we find the approximate area of the semicircle to be

$$A = \sum_{i=1}^{i=4} A_i = A_1 + A_2 + A_3 + A_4$$

$$= 3.48 + 7.48 + 7.48 + 3.48 = 21.92 \text{ in}^2 \quad Ans.$$

But this is a very poor approximation because the actual area is

$$A = \frac{1}{2} \pi r^2 = \frac{1}{2} \pi (4)^2 = 25.2 \text{ in}^2$$

It will be interesting to learn what happens to the accuracy if we use more rectangles. Accordingly, using the same process, we have divided the semicircle into eight rectangles, each 1 in wide (Fig. 9c). Summing the areas as before, we get

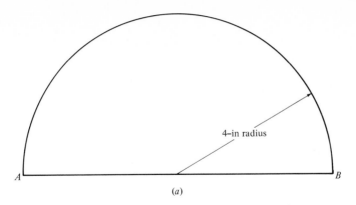

(a)

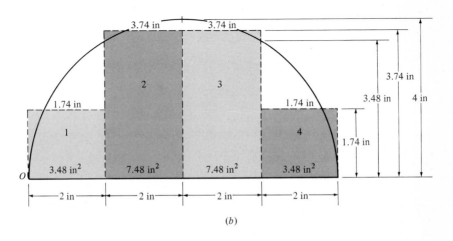

(b)

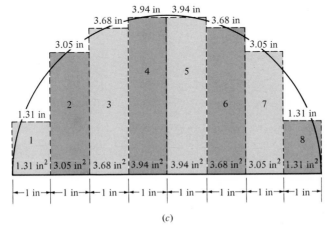

(c)

FIGURE 9. Finding the area of a semicircle by approximate summation.

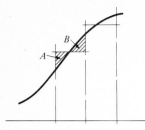

FIGURE 10. With the graphical method the triangular areas A and B should be equal.

$$A = \sum_{i=1}^{i=8} A_i = 1.31 + 3.05 + 3.68 + 3.94 + 3.68 + 3.05 + 1.31$$

$$= 23.96 \text{ in}^2 \quad Ans.$$

With four rectangles, the percentage error is

$$e = \frac{25.2 - 21.92}{25.2} (100) = 13\%$$

But with eight rectangles, the error is

$$e = \frac{25.2 - 23.96}{25.2} (100) = 4.9\%$$

which is a substantial improvement.

Sometimes in summing a continuous distribution of something, the rectangles are formed graphically and the height of each is adjusted by eye. In such cases, the errors can be minimized by proportioning the height so as to equalize the area of two small triangles, as shown in Fig. 10.

6. SUMMATION BY INTEGRATION Whenever the equation of the outline of an area is known or can be found, then a quick and easy method of getting the area is a process called *integration*. Integration can be roughly defined as the mathematical process of summing an infinitely large number of infinitesimal areas. This appears to the novice as the process of summing an infinitely large string of zeros yet, surprisingly, the exact answer is obtained. Before illustrating the procedure, it is necessary to build a small vocabulary.

Figure 11*a* is a graph of a line

$$y = f(x)$$

which is read "y is a function of x." The particular function is not specified in this instance; that is, it means just any old function. So it could be one of the following functions:

$$y = Ax^2 + B \quad y = -x + C\sqrt{x} \quad y = D \ln x$$

The lines in Fig. 11*a* labeled $x = a$ and $x = b$ are called the *boundaries* or the *limits* of the shaded area designated as A. The vertical line $x = a$ is often called the *lower boundary*. Then, the line $x = b$ is the *upper boundary*.

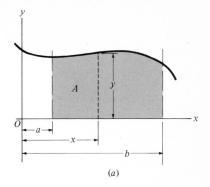

(a)

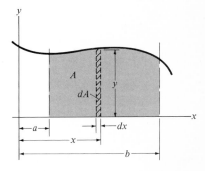

FIGURE 11.

In Fig. 11*b*, at a distance *x* from the origin *O*, is shown a narrow strip of area *y* inches high. This area is labeled as being *dx* inches wide. The symbol *dx* is called a *differential* by mathematicians. By this they mean an infinitesimally small bit of the distance *x*. The symbol *dx* is to be regarded as a single symbol representing a very small bit of *x*. *It does not* mean *d* multiplied by *x*. The symbol *dx* is read *d* of *x*, or *differential x*.

Also shown in Fig. 11*b* is the differential area *dA*. This is simply a very small bit of the area *A*.

Differential strips such as the one shown in Fig. 11*b* can always be treated as rectangles, even though the ends may be curved, because they are infinitesimally small. And so the area of the strip is the length times the width, or

$$dA = y \, dx$$

Integration is the mathematical process of adding up or summing the area of an infinite number of strips. The integration symbol is an elongated version of the English letter S, meaning a *sum*, and is written

$$A = \int dA$$

This means that the area *A* is simply the sum of all the differential areas *dA*. It has the same significance as the equation

$$A = \sum_{i=1}^{i=\infty} dA_i$$

But it is no longer necessary to label each particular strip because there are always an infinite number of them.

The expression, called a *definite integral*,

$$A = \int_a^b dA$$

means to sum or integrate beginning with the lower limit *a* and ending with the upper limit *b*.

EXAMPLE 1   Plot the curve

$$y = 1 + 2x - \frac{1}{2}x^2$$

and find the area bounded by this curve and the *x* axis between $x = 0$ and $x = 4$, between $x = 0$ and $x = 1$, and between $x = 1$ and $x = 4$.

SOLUTION

The given function is plotted in Fig. 12, and we first see that

$$dA = y \, dx$$

Substituting the value of *y* in this equation, we have

$$dA = \left(1 + 2x - \frac{x^2}{2}\right) dx$$

Therefore,

$$A = \int_a^b dA = \int_a^b \left(1 + 2x - \frac{x^2}{2}\right) dx$$

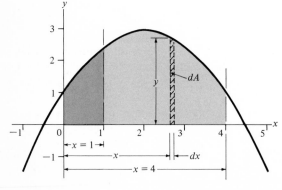

FIGURE 12.

where a and b are the lower and upper limits, respectively. Next, we factor the equation as follows:

$$A = \int_a^b dx + 2 \int_a^b x\, dx - \frac{1}{2} \int_a^b x^2\, dx$$

Now we use Table A-6 to find the integral of each term. The result is

$$A = x \Big|_a^b + 2 \left(\frac{x^2}{2}\right) \Big|_a^b - \frac{1}{2}\left(\frac{x^3}{3}\right)\Big|_a^b$$

This notation means that the term is to be evaluated between the two limits a and b. We can generalize this by writing

$$f(z)\Big|_a^b = f(b) - f(a)$$

Thus, the area between $x = 0$ and $x = 4$ is

$$A = x\Big|_0^4 + x^2\Big|_0^4 - \frac{x^3}{6}\Big|_0^4 = (4-0) + [(4)^2 - (0)^2] - \left[\frac{(4)^3}{6} - \frac{(0)^3}{6}\right]$$
$$= 4 + 16 - 10\tfrac{2}{3} = 9\tfrac{1}{3} \qquad Ans.$$

Similarly, we find the area between the ordinates $x = 0$ and $x = 1$ to be

$$A = x\Big|_0^1 + x^2\Big|_0^1 - \frac{x^3}{6}\Big|_0^1 = 1 + 1 - \tfrac{1}{6} = 1\tfrac{5}{6} \qquad Ans.$$

Then, the area from $x = 1$ to $x = 4$ must be the difference in these two, or

$$A = 9\tfrac{1}{3} - 1\tfrac{5}{6} = 7\tfrac{1}{2} \qquad Ans.$$

But this result can also be found as follows:

$$A = x\Big|_1^4 + x^2\Big|_1^4 - \frac{x^3}{6}\Big|_1^4 = (4-1) + [(4)^2 - (1)^2] - \left[\frac{(4)^3}{6} - \frac{(1)^3}{6}\right]$$
$$= 3 + 15 - 10\tfrac{1}{2} = 7\tfrac{1}{2} \qquad Ans. \qquad\qquad ////$$

EXAMPLE 2 Find the area of a right triangle whose base is 10 units long and whose altitude is 8 units.

SOLUTION

We begin by sketching the triangle and choosing a convenient coordinate system (Fig. 13). Next, we sketch an element of area dA at some distance x from the origin O. Designating the height of this element by h, we can use similar triangles to find

$$\frac{h}{10-x} = \frac{8}{10}$$

or $h = \dfrac{8}{10}(10-x) = 8 - \dfrac{8x}{10}$

Then $dA = h\,dx = \left(8 - \dfrac{8}{10}x\right)dx$

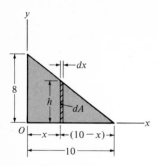

FIGURE 13.

We now have

$$A = \int_a^b dA = \int_0^{10} \left(8 - \frac{8}{10} x\right) dx$$

because we want to begin the integration at $x = 0$ and stop at $x = 10$, where the triangle ends.

Factoring again gives

$$A = 8 \int_0^{10} dx - \frac{8}{10} \int_0^{10} x \, dx$$

Using Table A-6, we find

$$A = 8x \Big|_0^{10} - \frac{8}{10} \frac{x^2}{2} \Big|_0^{10} = 8(10 - 0) - \frac{8}{10} \left[\frac{(10)^2}{2} - \frac{(0)^2}{2}\right]$$

$$= 80 - 40 = 40 \text{ square units} \qquad Ans.$$

Of course we could have used the formula for the area of a triangle and obtained the same result. ////

7. CENTROIDS OF AREAS BY INTEGRATION We can obtain another formula for determining centroidal distances merely by substituting integration signs for the summation signs in Eq. (2). This yields

$$\bar{x} = \frac{\int x' \, dA}{\int dA} = \frac{1}{A} \int x' \, dA \qquad (5)$$

where x' is the distance to the centroid of the element of area dA. By changing variables, we can get a corresponding formula for integration in the y direction. It is

$$\bar{y} = \frac{\int y' \, dA}{\int dA} = \frac{1}{A} \int y' \, dA \qquad (6)$$

EXAMPLE 3 Find $\bar{x}$ and $\bar{y}$ for the parabola of base b and height h in Fig. 14a.

SOLUTION

In Fig. 14b, we have chosen a vertical element of area. Since

$$dA = y \, dx = cx^2 \, dx$$

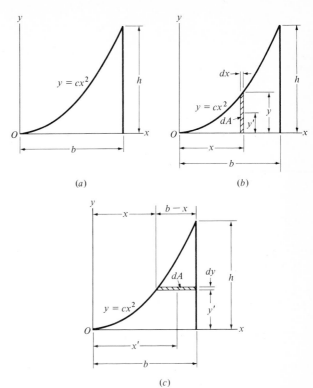

FIGURE 14.

we have

$$A = \int dA = c \int_0^b x^2 \, dx = c \frac{x^3}{3} \bigg|_0^b = \frac{cb^3}{3}$$

where Table A-6 has been used to evaluate the integral. Since $x' = x$ in Fig. 14b, Eq. (5) is written

$$\bar{x} = \frac{1}{A} \int x' \, dA = \frac{1}{A} \int_0^b (x)(cx^2 \, dx) = \frac{c}{A} \int_0^b x^3 \, dx$$

Upon evaluating this integral, we obtain

$$\bar{x} = \frac{cx^4}{4A} \bigg|_0^b = \frac{cb^4}{4A}$$

Since $A = cb^3/3$, we have

$$\bar{x} = \frac{3b}{4} \qquad Ans.$$

To find $\bar{y}$ using Fig. 14b, we write Eq. (6) as

$$\bar{y} = \frac{1}{A} \int y' \, dA = \frac{1}{A} \int_0^b \left(\frac{y}{2}\right)(cx^2 \, dx)$$

because $y' = y/2$. Note that we are to integrate along x, and so the upper limit is b, not h. Substituting the value of y and factoring gives

$$\bar{y} = \frac{1}{A} \int_0^b \left(\frac{cx^2}{2} \right) (cx^2 \, dx) = \frac{c^2}{2A} \int_0^b x^4 \, dx$$

Using the table of integrals, we get

$$\bar{y} = \frac{c^2}{2A} \frac{x^5}{5} \Big|_0^b = \frac{c^2 b^5}{10A}$$

Since $A = cb^3/3$ and $h = cb^2$, we have

$$\bar{y} = \frac{c^2 b^5}{10(cb^3/3)} = \frac{3h}{10} \qquad Ans.$$

The same results should, of course, be obtained when integrating from 0 to h using the horizontal strip of Fig. 14c. The applicable equations are

$$\bar{x} = \frac{1}{A} \int_0^h x' \, dA \qquad \bar{y} = \frac{1}{A} \int_0^h y' \, dA$$

In solving these equations, be sure to express x', y', A, and dA in terms of y and h.

////

EXAMPLE 4   Find the location of the centroid of a semicircular area.

SOLUTION

Sometimes a square element of area dA must be used instead of a rectangular strip. In such cases, double integration must be used. In this example, the double integration process is illustrated along with polar notation.

In Fig. 15, we select the x axis as the base of a semicircle of radius R, with the y axis passing vertically through the center. Then $\bar{x} = 0$ by symmetry, and we need only find $\bar{y}$. Thus, the equation to be solved is

$$\bar{y} = \frac{1}{A} \int y' \, dA$$

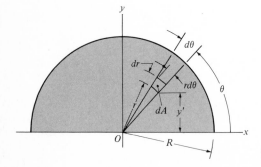

FIGURE 15.

We now locate an approximate element of area dA by the polar coordinates r and θ. This element has sides whose lengths are dr and $r\,d\theta$. Therefore,

$$dA = r\,dr\,d\theta$$

Since there are two differentials in this expression, we must integrate it twice, once with respect to θ and once with respect to r. This process is called *double integration*. It is written

$$A = dA = \int_0^\pi \int_0^R r\,dr\,d\theta$$

By placing parentheses in this expression, you can see better how the process works:

$$A = \int_0^\pi \left(\int_0^R r\,dr \right) d\theta$$

The inside operation should be performed first. Thus,

$$A = \int_0^\pi \left(\frac{r^2}{2} \bigg|_0^R \right) d\theta$$

Substituting the limits, we have

$$A = \int_0^\pi \frac{R^2}{2}\,d\theta$$

And integrating this gives

$$A = \frac{R^2}{2}\,\theta \bigg|_0^\pi = \frac{\pi R^2}{2}$$

which we recognize as the area of a semicircle. Of course, this is a well-known formula and it wouldn't have been necessary for us to develop it, but it does illustrate the double-integration process with a known result.

So now we have a formula for A and dA. From Fig. 15, the distance y' is

$$y' = r \sin \theta$$

and so the integral for $\bar{y}$ is written

$$\bar{y} = \frac{1}{A} \int_0^\pi \int_0^R (r \sin \theta)\,r\,dr\,d\theta$$

Performing the inside integration first gives

$$\bar{y} = \frac{1}{A} \int_0^\pi \left(\frac{r^3}{3} \sin \theta \right)_0^R d\theta$$

$$= \frac{1}{A} \int_0^\pi \frac{R^3}{3} \sin \theta\,d\theta$$

Then, with the help of a table of integrals, we find

$$\bar{y} = \frac{1}{A}\frac{R^3}{3}\left(-\cos\theta\right)_0^\pi = \frac{1}{A}\frac{R^3}{3}\left[-(\cos\pi - \cos 0)\right]$$

$$= \frac{1}{A}\frac{R^3}{3}\left[-(-1-1)\right] = \frac{2R^3}{3A}$$

Substituting $A = \pi R^2/2$ gives, finally,

$$\bar{y} = \frac{4R}{3\pi} \qquad Ans. \qquad\qquad ////$$

8. VOLUMES   Let some curve such as AB in Fig. 16 lie in the xy plane without intersecting the x axis, and let us revolve this curve about the x axis. In revolving about x, the curve AB generates a *surface of revolution*. There are two useful relations having to do with surfaces and volumes of revolution, called the *theorems of Pappus and Guldinus*, which we now wish to investigate.

The first theorem states that *the area of the surface of revolution is equal to the length of the generating line multiplied by the circumference of the circle traversed by the centroid of the generating line*.

To prove the first theorem, we select an element of length dl of the generating curve at a distance y from the x axis, as shown in Fig. 16. The area generated by this element in rotating about the x axis is

$$dA = (2\pi y)\ dl$$

Upon integrating both sides, we find the total area to be

$$A = 2\pi \int y\ dl = 2\pi \bar{y} l \qquad\qquad (7)$$

because, by definition,

$$\bar{y} = \frac{1}{l}\int y\ dl \qquad\qquad (8)$$

Equation (7) is a statement of the first theorem in algebraic form.

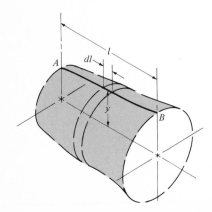

FIGURE 16.

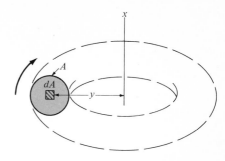

FIGURE 17.

Figure 17 illustrates a *volume of revolution*. Let some plane surface *A* lie in the *xy* plane. We select the *x* axis so that it does not intersect the boundary of the plane area, or, at most, so that the *x* axis is tangent to the boundary. A *volume of revolution* is now produced by revolving the plane area about the *x* axis.

The second theorem of Pappus and Guldinus states that *the volume of the body of revolution generated by the plane area is equal to the area of the plane surface multiplied by the circumference of the circle traversed by the centroid of the plane area.*

To prove the second theorem, we select an element of area *dA* at distance *y* from the *x* axis (Fig. 17). When this element is rotated about *x*, it generates a body of revolution whose volume is

$$dV = (2\pi y)\ dA$$

Upon integrating both sides, we find the total volume to be

$$V = 2\pi \int y\ dA = 2\pi \bar{y} A \tag{9}$$

Because, by definition, the centroid of an area is located at

$$\bar{y} = \frac{1}{A} \int y\ dA \tag{10}$$

second moments of areas

1. INTRODUCTION The expression

$$A\bar{x} = \int x'\, dA$$

which appeared as Eq. (5) in Appendix A-7 is a *first moment of an area* because an element of area dA is multiplied by the distance x' from some stated axis. A *second moment of an area* is obtained when the element of area is multiplied by the square of a distance to some stated axis. Thus, the expressions

$$\int x^2\, dA \qquad \int y^2\, dA \qquad \int r^2\, dA \tag{1}$$

are all second moments of areas.

It so happens that mathematical expressions for second moments of areas arise in the development of the formulas for the stresses in beams, columns, and shafts. These expressions could be permitted to remain in these formulas, but it is both convenient and customary to evaluate them separately. The expressions resemble the equation for *moment of inertia,* which is

$$\int \rho^2\, dm \tag{2}$$

where dm is an element of mass and ρ (Greek letter rho) is the distance to some stated axis. Because of the resemblance, Eqs. (1) are frequently called the equations for moment of inertia too, but this is a misnomer because an area cannot have inertia. Nevertheless, the practice is widespread and you must be alert to the differences. There will be no confusion if you use the terms *second moment of area* or *moment of inertia of an area* for Eq. (1), and *moment of inertia of a mass* for Eq. (2).

In this book we shall not require the expression for the moment of inertia of a mass, and so there should be no confusion.

2. RECTANGULAR MOMENT OF INERTIA In this section we shall deal with the specific problem of finding the *second moment of a plane area* about one of the rectangular axes. The formulas are

$$I_x = \int y^2\, dA \qquad I_y = \int x^2\, dA \tag{3}$$

In using these equations, the rule to be followed is that all parts of the element dA must be the same distance from the axis of moments.

EXAMPLE 1 Find the moment of inertia of a rectangular area b units wide and h units high about its centroidal axis, about its base, and about an axis h units away from its base.

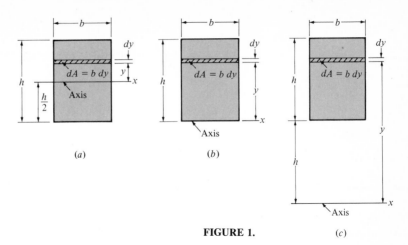

FIGURE 1.

(a)   (b)   (c)

SOLUTION

Figure 1*a* shows the rectangle with the *x* axis coincident with the centroidal axis $h/2$ units above the base. The area dA has been selected so that it is everywhere *y* units from *x*. The applicable equation is

$$I_x = \int y^2 \, dA$$

where the subscript *x* indicates that moments are to be taken about the *x* axis. The integration limits are from $y = -h/2$ to $y = +h/2$. As shown, the area is $dA = b \, dy$. Therefore, we have

$$I_x = \int_{-h/2}^{+h/2} y^2 (b \, dy) = \frac{by^3}{3} \bigg|_{-h/2}^{+h/2} = \frac{bh^3}{12} \qquad Ans.$$

For the second part of the problem, we select the axis of moments at the base of the rectangular area, as shown in Fig. 1*b*. The same equation applies, but now the limits are from $y = 0$ to $y = h$. We therefore have

$$I_x = \int y^2 \, dA = \int_0^h y^2 (b \, dy) = \frac{by^3}{3} \bigg|_0^h = \frac{bh^3}{3} \qquad Ans.$$

We note that the moment is four times as much as it is about the centroidal axis. For the third part (Fig. 1*c*), the limits of integration are from $y = h$ to $y = 2h$. Therefore,

$$I_x = \frac{by^3}{3} \bigg|_h^{2h} = \frac{7bh^3}{3} \qquad Ans. \qquad ////$$

EXAMPLE 2   A right triangle has a base of width *b* and an altitude *h*. Find the second moment of area of the triangle about an axis through the base.

SOLUTION

Referring to Fig. 2, the applicable equation is

$$I_x = \int y^2 \, dA$$

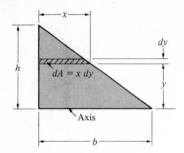

FIGURE 2.

because the axis x has been selected coincident with the base. The limits of integration are seen to be from $y = 0$ to $y = h$. Before integrating, it is necessary to express the element length x in terms of y because, as shown, $dA = x\,dy$. We can do this by noting that the triangle whose base is x is similar to the entire triangle. Therefore,

$$\frac{x}{b} = \frac{h - y}{h}$$

and so

$$x = \frac{b}{h}\,(h - y)$$

The integral then becomes

$$I_x = \int y^2\,dA = \int y^2(x\,dy) = \frac{b}{h}\int_0^h y^2(h - y)\,dy$$

This integral may be factored and then evaluated as follows:

$$I_x = \frac{b}{h}\left(h\int_0^h y^2\,dy - \int_0^h y^3\,dy\right)$$

$$= \frac{b}{h}\left[h\left(\frac{y^3}{3}\right)_0^h - \left(\frac{y^4}{4}\right)_0^h \right] = \frac{bh^3}{12} \qquad Ans. \qquad\qquad ////$$

3. POLAR MOMENT OF INERTIA In the previous section we learned that a rectangular moment of inertia of an area is found by taking the second moment about an axis in the *same* plane as the area. In contrast, the *polar moment of inertia of an area* is the second moment of the area about an axis *normal to the plane* of the area. The equation is

$$J = \int \rho^2\,dA \tag{4}$$

where ρ is the distance from the axis of moments to the element of area dA.

EXAMPLE 3 Find the polar moment of inertia of the area of a circle about its centroidal axis.

SOLUTION

In Fig. 3 we select a circular element of area dA of thickness $d\rho$ and at a radius ρ

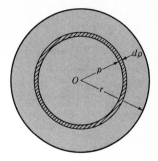

FIGURE 3.

from the moment axis O. Then the differential area is the product of the thickness $d\rho$ times the circumference $2\pi\rho$, or

$$dA = 2\pi\rho \, d\rho$$

Equation (4) now becomes

$$J = \int \rho^2 \, dA = 2\pi \int_0^r \rho^3 \, d\rho = 2\pi \left. \frac{\rho^4}{4} \right|_0^r = \frac{\pi r^4}{2} \qquad Ans.$$

Or, in terms of the diameter, the answer is

$$J = \frac{\pi(d/2)^4}{2} = \frac{\pi d^4}{32} \qquad Ans. \qquad\qquad ////$$

4. RADIUS OF GYRATION   Since area moment of inertia can be considered as the product of a distance squared with an area, its units are units of length raised to the fourth power. Thus, in the English system, the units of I and J would normally be in^4, and in SI, m^4.

We can also think of area moment of inertia as the total area times the square of a fictitious distance. In this form, we could write

$$I_x = \int y^2 \, dA = k_x^2 A \qquad\qquad (5)$$

Or, in polar form,

$$J_z = \int \rho^2 \, dA = k_z^2 A \qquad\qquad (6)$$

where A is the total area and k is a fictitious distance that is called the *radius of gyration*. The subscript of k designates the axis about which it is taken.

If we think of the total area as being concentrated along a line, then radius of gyration is the distance from the moment axis to the line.

5. TRANSFORMATION METHODS   Assume that the moment of inertia of an area about its own centroidal axis is known. Assume also that we wish to compute the moment of inertia of the area about another axis which is to be parallel to the centroidal axis. In

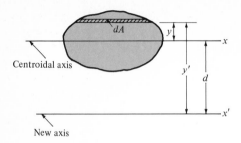

FIGURE 4.

this section we present a formula for computing the moment of inertia about the parallel axis without using integration.

Suppose, in Fig. 4, that the x axis passes through the centroid of the area A. Then

$$I = I_x = \int y^2 \, dA$$

In Fig. 4 the axis x' is parallel to x and removed a distance d. What is the moment of inertia I' about the x' axis? If we were to use integration, the moment of inertia would be

$$I' = \int y'^2 \, dA$$

But since $y' = y + d$, we have

$$y'^2 = y^2 + 2yd + d^2$$

Therefore,

$$I' = \int y^2 \, dA + 2d \int y \, dA + d^2 \int dA$$

In this equation,

$$\int y^2 \, dA = I_G$$

where the subscript G is used to indicate the moment of inertia is about the centroidal axis of the section. Also,

$$\int y \, dA = \bar{y}A = 0$$

because the x axis passes through the centroidal axis of the section. Then

$$d^2 \int dA = d^2 A$$

And so

$$I' = I_G + d^2 A \tag{7}$$

which is called the *transfer formula*. Here, I_G is the moment of inertia about the centroidal axis of the section and I' is the moment of inertia about another axis parallel to the centroidal axis and at a distance d from it. Using this formula and the worked-out second moments from Table A-15, for example, makes it possible to compute the area moments of inertia for a large variety of sections, as we shall learn in the next section.

6. APPLICATIONS   Using the transfer formula [Eq. (7)], together with some of the formulas previously developed, we show in this section methods of computing second moments of inertias of a variety of geometries by means of examples.

EXAMPLE 4   Find the area moment of inertia of a right triangle whose base is 6 in and whose altitude is 9 in about a centroidal axis parallel to the base and about an axis through the apex parallel to the base.

SOLUTION

Referring to Fig. 5, let us designate the various moments of inertia as I_G, I_1, and I_2, where the subscripts refer to axes through the centroid, through the base, and through the apex, respectively, as shown. Then, from Example 2,

$$I_1 = \frac{bh^3}{12} = \frac{(6)(9)^3}{12} = 364.5 \text{ in}^4 \qquad Ans.$$

Next, using the transfer formula, we learn that

$$I_1 = I_G + d^2 A$$

and so

$$I_G = I_1 - d^2 A$$

Since $A = bh/2$ and $d^2 = (h/3)^2 = h^2/9$, we have

$$I_G = \frac{bh^3}{12} - \frac{h^2}{9}\frac{bh}{2} = \frac{bh^3}{36}$$

Therefore,

$$I_G = \frac{(6)(9)^3}{36} = 121.5 \text{ in}^4 \qquad Ans.$$

Using the transfer formula again to get the moment of inertia about an axis through the apex gives

$$I_2 = I_G + d^2 A = \frac{bh^3}{36} + \left(\frac{2h}{3}\right)^2 \left(\frac{bh}{2}\right) = \frac{bh^3}{4}$$

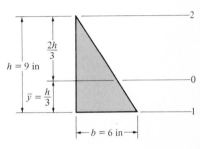

FIGURE 5.

And so

$$I_2 = \frac{(6)(9)^3}{4} = 1093.5 \text{ in}^4 \quad Ans.$$

////

EXAMPLE 5   Locate the centroid of the T section shown in Fig. 6 and then compute the area moment of inertia of the T about an axis through the centroid parallel to the base.

SOLUTION

We first divide the T into the two rectangular areas a and b, as shown in Fig. 7a, designating each centroid as G_a and G_b. Then the areas are

$$A_a = (1)(6) = 6 \text{ in}^2 \quad A_b = (4)(1.5) = 6 \text{ in}^2$$

Next, taking *first* moments about the x axis gives

$$A\bar{y} = \bar{y}_a A_a + \bar{y}_b A_b$$

and so

$$\bar{y} = \frac{(4.5)(6) + (2)(6)}{12} = 3.25 \text{ in} \quad Ans.$$

The moments of inertia of each rectangular area about horizontal axes through their own centroids are

$$I_a = \frac{bh^3}{12} = \frac{(6)(1)^3}{12} = 0.50 \text{ in}^4$$

$$I_b = \frac{bh^3}{12} = \frac{(1.5)(4)^3}{12} = 8 \text{ in}^4$$

Next, refer to Fig. 7b where the centroid G of the T is seen to be 1.25 in from G_a and 1.25 in from G_b. The area moment of inertia of the T about a horizontal axis through G is found by applying the transfer formula twice, once for each rectangle, as follows:

$$I = (I_a + d_a^2 A_a) + (I_b + d_b^2 A_b)$$
$$= [0.50 + (1.25)^2(6)] + [8 + (1.25)^2(6)] = 27.25 \text{ in}^4 \quad Ans.$$

////

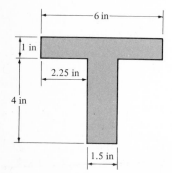

FIGURE 6.

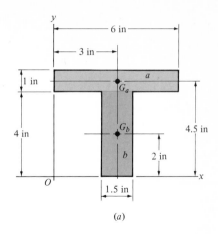

(a)

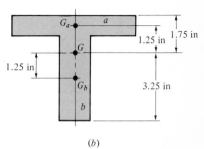

(b)

FIGURE 7.

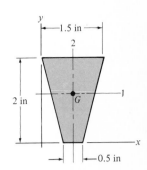

FIGURE 8.

EXAMPLE 6   Locate the centroid G of the trapezoidal cross section shown in Fig. 8 and find the area moments of inertia about axes 1 and 2.

SOLUTION

In the horizontal direction G is located by symmetry 0.75 in from y. To obtain $\bar{y}$, we visualize the trapezoid as composed of a positive rectangular area and two nega-

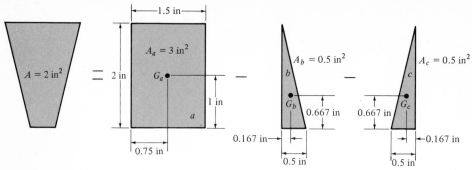

FIGURE 9.

tive triangular areas, as illustrated in Fig. 9. This figure also shows the areas of each segment and the locations of the individual centroids. To find $\bar{y}$, we write

$$\bar{y}A = \bar{y}_a A_a - \bar{y}_b A_b - \bar{y}_c A_c$$

Then

$$\bar{y} = \frac{1}{A}(\bar{y}_a A_a - \bar{y}_b A_b - \bar{y}_c A_c)$$

$$= \frac{1}{2}[(1)(3) - (0.667)(0.5) - (0.667)(0.5)]$$

$$= 1.167 \text{ in} \qquad Ans.$$

The next step is illustrated in Fig. 10. First we must find the moments of inertia of the three segments about axes through their own centroids parallel to axis 1. These are

$$I_a = \frac{bh^3}{12} = \frac{(1.5)(2)^3}{12} = 1 \text{ in}^4$$

$$I_b = I_c = \frac{bh^3}{36} = \frac{(0.5)(2)^3}{36} = 0.111 \text{ in}^4$$

Incorporating the transfer formula, the moment of inertia of the trapezoid about its own centroid will be

$$I_1 = (I_a + d_a^2 A_a) - (I_b + d_b^2 A_b) - (I_c + d_c^2 A_c)$$

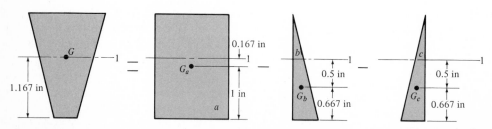

FIGURE 10.

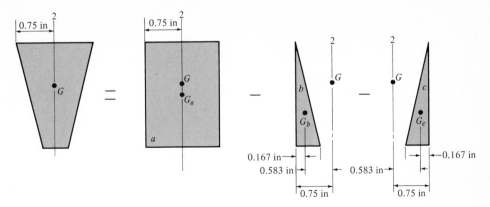

FIGURE 11.

Upon substituting the appropriate values as shown in Fig. 10, we get

$$I_1 = [1 + (0.167)^2(3)] - [0.111 + (0.5)^2(0.5)] - [0.111 + (0.5)^2(0.5)]$$
$$= 0.612 \text{ in}^4 \quad Ans.$$

A similar procedure is followed to get I_2. Referring to Fig. 11, we first find the moments of inertia of the three segments about vertical axes through their own centroids. These are

$$I_a = \frac{hb^3}{12} = \frac{(2)(1.5)^3}{12} = 0.562 \text{ in}^4$$

$$I_b = I_c = \frac{hb^3}{36} = \frac{(2)(0.5)^3}{36} = 0.0069 \text{ in}^4$$

We then compute I_2 as

$$I_2 = I_a - (I_b + d_b^2 A_b) - (I_c + d_c^2 A_c)$$
$$= 0.562 - [0.0069 + (0.583)^2(0.5)] - [0.0069 + (0.583)^2(0.5)]$$
$$= 0.209 \text{ in}^4 \quad Ans. \qquad ////$$

7. PRINCIPAL AXES   Sometimes we may encounter the integral

$$I_{xy} = \int xy \, dA \qquad (8)$$

which is called the *product of inertia*. In evaluating this integral (Fig. 12), each element of area dA must be multiplied by the product xy of its coordinates and the integration carried out over the entire area. The product of inertia can be either positive or negative because x and y can have positive or negative values.

Let us suppose that one of the axes, say the y axis, is an axis of symmetry. Then for every element of area dA located by a positive x, there will exist a twin element, symmetrically located, having a corresponding negative x. These will sum to zero in the integration, and hence the product of inertia is always zero when either x or y is an axis of symmetry.

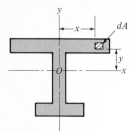

FIGURE 12.

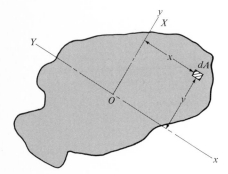

FIGURE 13.

Now consider Fig. 13, in which the product of inertia of the figure, in terms of the xy coordinate system, will be given by the equation

$$I_{xy} = \int xy \, dA$$

Now choose an XY coordinate system with origin at the same point O, but with $X = y$ and $Y = -x$ as shown. In this system, the product of inertia is

$$I_{XY} = \int XY \, dA = \int -yx \, dA = -I_{xy}$$

In words, we rotated the xy system 90° to get the XY system. In doing so, the product of inertia changed from $+I_{xy}$ to $-I_{xy}$. This product is a continuous function and so it must have passed through zero somewhere during the rotation. The two axes corresponding to this zero position are called the *principal axes* of the section. When the centroid of the plane area is taken as the origin of the axes, with one of the axes being an axis of symmetry, then these two axes are the *centroidal principal axes*.

It is not difficult to find the location of the principal axes by mathematical techniques when I_x, I_y, and I_{xy} are given. Since most of the areas with which we shall deal do have an axis of symmetry, the principal axes are readily located, and so we shall not develop this technique here.

TABLE A-9
Physical constants of materials.[1]

material	modulus of elasticity, E psi	modulus of elasticity, E Pa	modulus of rigidity, G psi	modulus of rigidity, G Pa	Poisson's ratio, μ	units of weight, w lbf/in³	units of weight, w lbf/ft	units of weight, w kN/m³
Aluminum (all alloys)	1.03 E+07	7.10 E+10	3.80 E+06	2.62 E+10	0.334	0.098	169	26.6
Brass	1.54 E+07	1.06 E+11	5.82 E+06	4.01 E+10	0.324	0.309	534	83.8
Carbon steel	3.0 E+07	2.07 E+11	1.15 E+07	7.93 E+10	0.292	0.282	487	76.5
Cast iron, gray	1.45 E+07	1.00 E+11	6.0 E+06	4.14 E+10	0.211	0.260	450	70.6
Concrete	3.0 E+06	2.07 E+10	1.4 E+06	9.65 E+09	0.1	0.087	150	23.6
Copper	1.72 E+07	1.19 E+11	6.49 E+06	4.47 E+10	0.326	0.322	556	87.3
Douglas fir	1.6 E+06	1.1 E+10	6.0 E+05	4.14 E+09	0.33	0.016	27.7	4.34
Glass	6.7 E+06	4.62 E+10	2.7 E+06	1.86 E+10	0.245	0.094	162	25.4
Lead	5.3 E+06	3.65 E+10	1.9 E+06	1.31 E+10	0.425	0.411	710	111.5
Longleaf pine	1.99 E+06	1.37 E+10				0.021	36.2	5.68
Magnesium	6.5 E+06	4.48 E+10	2.4 E+06	1.65 E+10	0.350	0.065	112	17.6
Shortleaf pine	1.76 E+06	1.21 E+10				0.018	31.8	4.99
Stainless steel	2.76 E+07	1.90 E+11	1.06 E+07	7.31 E+10	0.305	0.280	484	76.0
White oak	1.62 E+06	1.12 E+10				0.024	41.8	6.56
White pine	1.28 E+06	8.82 E+09				0.013	22.5	3.53

[1] These are average properties that have been compiled from a number of sources; do not use these values for actual design without first obtaining experimental verification.

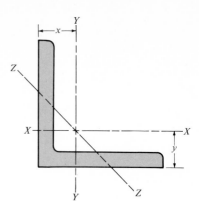

TABLE A-10
Properties of rolled-steel angles – equal legs.[1]

size in	thickness in	weight lbf/ft	area in²	axis X-X and axis Y-Y				axis Z-Z
				I in⁴	I/c in³	k in	x or $\bar{y}$ in	k in
8×8	1	51.0	15.00	89.0	15.8	2.44	2.37	1.56
	$\frac{3}{4}$	38.9	11.44	69.7	12.2	2.47	2.28	1.57
	$\frac{1}{2}$	26.4	7.75	48.6	8.4	2.50	2.19	1.59
6×6	1	37.4	11.0	35.5	8.6	1.80	1.86	1.17
	$\frac{3}{4}$	28.7	8.44	28.2	6.7	1.83	1.78	1.17
	$\frac{1}{2}$	19.6	5.75	19.9	4.6	1.86	1.68	1.18
	$\frac{3}{8}$	14.9	4.36	15.4	3.5	1.88	1.64	1.19
5×5	$\frac{3}{4}$	23.6	6.94	15.7	4.5	1.51	1.52	0.97
	$\frac{1}{2}$	16.2	4.75	11.3	3.2	1.54	1.43	0.98
	$\frac{3}{8}$	12.3	3.61	8.8	2.4	1.56	1.39	0.99
4×4	$\frac{3}{4}$	18.5	5.44	7.7	2.8	1.19	1.27	0.78
	$\frac{1}{2}$	12.8	3.75	5.6	2.0	1.22	1.18	0.78
	$\frac{3}{8}$	9.8	2.86	4.4	1.5	1.23	1.14	0.79
	$\frac{1}{4}$	6.6	1.94	3.0	1.1	1.25	1.09	0.80
$3\frac{1}{2} \times 3\frac{1}{2}$	$\frac{1}{2}$	11.1	3.25	3.6	1.5	1.06	1.06	0.68
	$\frac{3}{8}$	8.5	2.48	2.9	1.2	1.07	1.01	0.69
	$\frac{1}{4}$	5.8	1.69	2.0	0.79	1.09	0.97	0.69
3×3	$\frac{1}{2}$	9.4	2.75	2.2	1.1	0.90	0.93	0.58
	$\frac{3}{8}$	7.2	2.11	1.8	0.83	0.91	0.89	0.58
	$\frac{1}{4}$	4.9	1.44	1.2	0.58	0.93	0.84	0.59
$2\frac{1}{2} \times 2\frac{1}{2}$	$\frac{1}{2}$	7.7	2.25	1.2	0.72	0.74	0.81	0.49
	$\frac{3}{8}$	5.9	1.73	0.98	0.57	0.75	0.76	0.49
	$\frac{1}{4}$	4.1	1.19	0.70	0.39	0.77	0.72	0.49
2×2	$\frac{3}{8}$	4.7	1.36	0.48	0.35	0.59	0.64	0.39
	$\frac{1}{4}$	3.19	0.94	0.35	0.25	0.61	0.59	0.39
	$\frac{1}{8}$	1.65	0.48	0.19	0.13	0.63	0.55	0.40
$1\frac{3}{4} \times 1\frac{3}{4}$	$\frac{1}{4}$	2.77	0.81	0.23	0.19	0.53	0.53	0.34
	$\frac{1}{8}$	1.44	0.42	0.13	0.10	0.55	0.48	0.35
$1\frac{1}{2} \times 1\frac{1}{2}$	$\frac{1}{4}$	2.34	0.69	0.14	0.13	0.45	0.47	0.29
	$\frac{3}{16}$	1.80	0.53	0.11	0.10	0.46	0.44	0.29
	$\frac{1}{8}$	1.23	0.36	0.08	0.07	0.47	0.42	0.30

TABLE A-10
Properties of rolled-steel angles — equal legs (continued).[1]

size in	thickness in	weight lbf/ft	area in²	axis X-X and axis Y-Y				axis Z-Z
				I in⁴	I/c in³	k in	x or $\overline{y}$ in	k in
$1\frac{1}{4} \times 1\frac{1}{4}$	$\frac{1}{4}$	1.92	0.56	0.08	0.09	0.37	0.40	0.24
	$\frac{3}{16}$	1.48	0.43	0.06	0.07	0.38	0.38	0.24
	$\frac{1}{8}$	1.01	0.30	0.04	0.05	0.38	0.36	0.25
1×1	$\frac{1}{4}$	1.49	0.44	0.04	0.06	0.29	0.34	0.20
	$\frac{3}{16}$	1.16	0.34	0.03	0.04	0.30	0.32	0.19
	$\frac{1}{8}$	0.80	0.23	0.02	0.03	0.30	0.30	0.20

[1] For additional sizes, see the current volume of "Steel Construction," American Institute of Steel Construction (AISC), New York.

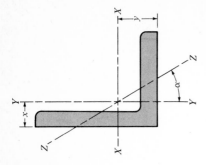

TABLE A-11
Properties of rolled-steel angles — unequal legs.[1]

size in	thickness in	weight lbf/ft	area in²	axis X-X I in⁴	axis X-X I/c in³	axis X-X k in	axis X-X $\bar{y}$ in	axis Y-Y I in⁴	axis Y-Y I/c in³	axis Y-Y k in	axis Y-Y $\bar{x}$ in	axis Z-Z k in	axis Z-Z tan α
8 × 6	1	44.2	13.00	80.8	15.1	2.49	2.65	38.8	8.9	1.73	1.65	1.28	0.543
	3/4	33.8	9.94	63.4	11.7	2.53	2.56	30.7	6.9	1.76	1.56	1.29	0.551
	1/2	23.0	6.75	44.3	8.0	2.56	2.47	21.7	4.8	1.79	1.47	1.30	0.558
8 × 4	1	37.4	11.00	69.6	14.1	2.52	3.05	11.6	3.9	1.03	1.05	0.85	0.247
	3/4	28.7	8.44	54.9	10.9	2.55	2.95	9.4	3.1	1.05	0.95	0.85	0.258
	1/2	19.6	5.75	38.5	7.5	2.59	2.86	6.7	2.2	1.08	0.86	0.86	0.267
6 × 4	3/4	23.6	6.94	24.5	6.3	1.88	2.08	8.7	3.0	1.12	1.08	0.86	0.428
	1/2	16.2	4.75	17.4	4.3	1.91	1.99	6.3	2.1	1.15	0.99	0.87	0.440
	3/8	12.3	3.61	13.5	3.3	1.93	1.94	4.9	1.6	1.17	0.94	0.88	0.446
5 × 3	1/2	12.8	3.75	9.5	2.9	1.59	1.75	2.6	1.1	0.83	0.75	0.65	0.357
	3/8	9.8	2.86	7.4	2.2	1.61	1.70	2.0	0.89	0.84	0.70	0.65	0.364
	1/4	6.6	1.94	5.1	1.5	1.62	1.66	1.4	0.61	0.86	0.66	0.66	0.371

Size	Thickness											
4 × 3½	½	11.9	3.50	5.3	1.9	1.23	1.25	3.8	1.5	1.04	1.00	0.72	0.750
	⅜	9.1	2.67	4.2	1.5	1.25	1.21	3.0	1.2	1.06	0.96	0.73	0.755
	¼	6.2	1.81	2.9	1.0	1.27	1.16	2.1	0.81	1.07	0.91	0.73	0.759
4 × 3	⅝	13.6	3.98	6.0	2.3	1.23	1.37	2.9	1.4	0.85	0.87	0.64	0.534
	⅜	8.5	2.48	4.0	1.5	1.26	1.28	1.9	0.87	0.88	0.78	0.64	0.551
	¼	5.8	1.69	2.8	1.0	1.28	1.24	1.4	0.60	0.90	0.74	0.65	0.558
3 × 2½	½	8.5	2.50	2.1	1.0	0.91	1.00	1.3	0.74	0.72	0.75	0.52	0.667
	⅜	6.6	1.92	1.7	0.81	0.93	0.96	1.0	0.58	0.74	0.71	0.52	0.676
	¼	4.5	1.31	1.2	0.56	0.95	0.91	0.74	0.40	0.75	0.66	0.53	0.684
3 × 2	½	7.7	2.25	1.9	1.0	0.92	1.08	0.67	0.47	0.55	0.58	0.43	0.414
	⅜	5.9	1.73	1.5	0.78	0.94	1.04	0.54	0.37	0.56	0.54	0.43	0.428
	¼	4.1	1.19	1.1	0.54	0.95	0.99	0.39	0.26	0.57	0.49	0.43	0.440
2½ × 2	⅜	5.3	1.55	0.91	0.55	0.77	0.83	0.51	0.36	0.58	0.58	0.42	0.614
	¼	3.62	1.06	0.65	0.38	0.78	0.79	0.37	0.25	0.59	0.54	0.42	0.626
2½ × 1½	⅜	4.7	1.36	0.82	0.52	0.78	0.92	0.22	0.20	0.40	0.42	0.32	0.340
	¼	3.19	0.94	0.59	0.36	0.79	0.88	0.16	0.14	0.41	0.38	0.32	0.357
2 × 1½	¼	2.77	0.81	0.32	0.24	0.62	0.66	0.15	0.14	0.43	0.41	0.32	0.543
	3/16	2.12	0.62	0.25	0.18	0.63	0.64	0.12	0.11	0.44	0.39	0.32	0.551
	⅛	1.44	0.42	0.17	0.13	0.64	0.62	0.09	0.08	0.45	0.37	0.33	0.558
1¾ × 1¼	¼	2.34	0.69	0.20	0.18	0.54	0.60	0.09	0.10	0.35	0.35	0.27	0.486
	3/16	1.80	0.53	0.16	0.14	0.55	0.58	0.07	0.08	0.36	0.33	0.27	0.496
	⅛	1.23	0.36	0.11	0.09	0.56	0.56	0.05	0.05	0.37	0.31	0.27	0.506

[1] For additional sizes, see the current volume of "Steel Construction," American Institute of Steel Construction (AISC), New York.

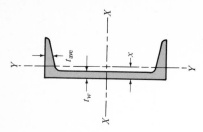

TABLE A-12
Properties of rolled-steel channels.

nominal size in	weight lbf/ft	area in²	depth in	flange width in	t_{avg} in	web t_w in	axis X-X I in⁴	I/c in³	k in	axis Y-Y I in⁴	I/c in³	k in	$\bar{x}$ in
18 × 4	58.0	16.98	18.00	4.200	0.625	0.700	670.7	74.5	6.29	18.5	5.6	1.04	0.88
	51.9	15.18	18.00	4.100	0.625	0.600	622.1	69.1	6.40	17.1	5.3	1.06	0.87
	45.8	13.38	18.00	4.000	0.625	0.500	573.5	63.7	6.55	15.8	5.1	1.09	0.89
	42.7	12.48	18.00	3.950	0.625	0.450	549.2	61.0	6.64	15.0	4.9	1.10	0.90
15 × 3⅜	50.0	14.64	15.00	3.716	0.650	0.716	401.4	53.6	5.24	11.2	3.8	0.87	0.80
	40.0	11.70	15.00	3.520	0.650	0.520	346.3	46.2	5.44	9.3	3.4	0.89	0.78
	33.9	9.90	15.00	3.400	0.650	0.400	312.6	41.7	5.62	8.2	3.2	0.91	0.79
12 × 3	30.0	8.79	12.00	3.170	0.501	0.510	161.2	26.9	4.28	5.2	2.1	0.77	0.68
	25.0	7.32	12.00	3.047	0.501	0.387	143.5	23.9	4.43	4.5	1.9	0.79	0.68
	20.7	6.03	12.00	2.940	0.501	0.280	128.1	21.4	4.61	3.9	1.7	0.81	0.70
10 × 2⅝	30.0	8.80	10.00	3.033	0.436	0.673	103.0	20.6	3.42	4.0	1.7	0.67	0.65
	25.0	7.33	10.00	2.886	0.436	0.526	90.7	18.1	3.52	3.4	1.5	0.68	0.62
	20.0	5.86	10.00	2.739	0.436	0.379	78.5	15.7	3.66	2.8	1.3	0.70	0.61
	15.3	4.47	10.00	2.600	0.436	0.240	66.9	13.4	3.87	2.3	1.2	0.72	0.64

$9 \times 2\frac{1}{2}$	20.0	5.86	9.00	2.648	0.413	0.448	60.6	13.5	3.22	2.4	1.2	0.65	0.59
	15.00	4.39	9.00	2.485	0.413	0.285	50.7	11.3	3.40	1.9	1.0	0.67	0.59
	13.4	3.89	9.00	2.430	0.413	0.230	47.3	10.5	3.49	1.8	0.97	0.67	0.61
$8 \times 2\frac{1}{4}$	18.75	5.49	8.00	2.527	0.390	0.487	43.7	10.9	2.82	2.0	1.0	0.60	0.57
	13.75	4.02	8.00	2.343	0.390	0.303	35.8	9.0	2.99	1.5	0.86	0.62	0.56
	11.5	3.36	8.00	2.260	0.390	0.220	32.3	8.1	3.10	1.3	0.79	0.63	0.58
$7 \times 2\frac{1}{8}$	14.75	4.32	7.00	2.299	0.366	0.419	27.1	7.7	2.51	1.4	0.79	0.57	0.53
	12.25	3.58	7.00	2.194	0.366	0.314	24.1	6.9	2.59	1.2	0.71	0.58	0.53
	9.8	2.85	7.00	2.090	0.366	0.210	21.1	6.0	2.72	0.98	0.63	0.59	0.55
6×2	13.0	3.81	6.00	2.157	0.343	0.437	17.3	5.8	2.13	1.1	0.65	0.53	0.52
	10.5	3.07	6.00	2.034	0.343	0.314	15.1	5.0	2.22	0.87	0.57	0.53	0.50
	8.2	2.39	6.00	1.920	0.343	0.200	13.0	4.3	2.34	0.70	0.50	0.54	0.52
$5 \times 1\frac{3}{4}$	9.0	2.63	5.00	1.885	0.320	0.325	8.8	3.5	1.83	0.64	0.45	0.49	0.48
	6.7	1.95	5.00	1.75	0.320	0.190	7.4	3.0	1.95	0.48	0.38	0.50	0.49
$4 \times 1\frac{5}{8}$	7.25	2.12	4.00	1.720	0.296	0.320	4.5	2.3	1.47	0.44	0.35	0.46	0.46
	5.4	1.56	4.00	1.580	0.296	0.180	3.8	1.9	1.56	0.32	0.29	0.45	0.46
$3 \times 1\frac{1}{2}$	6.0	1.75	3.00	1.596	0.273	0.356	2.1	1.4	1.08	0.31	0.27	0.42	0.46
	5.0	1.46	3.00	1.498	0.273	0.258	1.8	1.2	1.12	0.25	0.24	0.41	0.44
	4.1	1.19	3.00	1.410	0.273	0.170	1.6	1.1	1.17	0.20	0.21	0.41	0.44

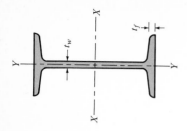

TABLE A-13
Properties of rolled-steel I beams.

nominal size in	weight lbf/ft	area in²	depth in	flange width in	t_f in	web t_w in	axis X-X I in⁴	I/c in³	k in	axis Y-Y I in⁴	I/c in³	k in
27 × 7⅞	120.0	35.13	24.00	8.048	1.102	0.798	3010.8	250.9	9.26	84.9	21.1	1.56
	105.9	30.98	24.00	7.875	1.102	0.625	2811.5	234.3	9.53	78.9	20.0	1.60
24 × 7	100.0	29.25	24.00	7.247	0.871	0.747	2371.8	197.6	9.05	48.4	13.4	1.29
	90.0	26.30	24.00	7.124	0.871	0.624	2230.1	185.8	9.21	45.5	12.8	1.32
	79.9	23.33	24.00	7.000	0.871	0.500	2087.2	173.9	9.46	42.9	12.2	1.36
20 × 7	95.0	27.74	20.00	7.200	0.916	0.800	1599.7	160.0	7.59	50.5	14.0	1.35
	85.0	24.80	20.00	7.053	0.916	0.653	1501.7	150.2	7.78	47.0	13.3	1.38
20 × 6¼	75.0	21.90	20.00	6.391	0.789	0.641	1263.5	126.3	7.60	30.1	9.4	1.17
	65.4	19.08	20.00	6.250	0.789	0.500	1169.5	116.9	7.83	27.9	8.9	1.21
18 × 6	70.0	20.46	18.00	6.251	0.691	0.711	917.5	101.9	6.70	24.5	7.8	1.09
	54.7	15.94	18.00	6.000	0.691	0.460	795.5	88.4	7.07	21.2	7.1	1.15
15 × 5½	50.0	14.59	15.00	5.640	0.622	0.550	481.1	64.2	5.74	16.0	5.7	1.05
	42.9	12.49	15.00	5.500	0.622	0.410	441.8	58.9	5.95	14.6	5.3	1.08

12 × 5¼	50.0 / 40.8	14.57 / 11.84	12.00 / 12.00	5.477 / 5.250	0.659 / 0.659	0.687 / 0.460	301.6 / 268.9	50.3 / 44.8	4.55 / 4.77	16.0 / 13.8	5.8 / 5.3	1.05 / 1.08
12 × 5	35.0 / 31.8	10.20 / 9.26	12.00 / 12.00	5.078 / 5.000	0.544 / 0.544	0.428 / 0.350	227.0 / 215.8	37.8 / 36.0	4.72 / 4.83	10.0 / 9.5	3.9 / 3.8	0.99 / 1.01
10 × 4⅝	35.0 / 25.4	10.22 / 7.38	10.00 / 10.00	4.944 / 4.660	0.491 / 0.491	0.594 / 0.310	145.8 / 122.1	29.2 / 24.4	3.78 / 4.07	8.5 / 6.9	3.4 / 3.0	0.91 / 0.97
8 × 4	23.0 / 18.4	6.71 / 5.34	8.00 / 8.00	4.171 / 4.000	0.425 / 0.425	0.441 / 0.270	64.2 / 56.9	16.0 / 14.2	3.09 / 3.26	4.4 / 3.8	2.1 / 1.9	0.81 / 0.84
7 × 3⅝	20.0 / 15.3	5.83 / 4.43	7.00 / 7.00	3.860 / 3.660	0.392 / 0.392	0.450 / 0.250	41.9 / 36.2	12.0 / 10.4	2.68 / 2.86	3.1 / 2.7	1.6 / 1.5	0.74 / 0.78
6 × 3⅜	17.25 / 12.5	5.02 / 3.61	6.00 / 6.00	3.565 / 3.330	0.359 / 0.359	0.465 / 0.230	26.0 / 21.8	8.7 / 7.3	2.28 / 2.46	2.3 / 1.8	1.3 / 1.1	0.68 / 0.72
5 × 3	14.75 / 10.0	4.29 / 2.87	5.00 / 5.00	3.284 / 3.000	0.326 / 0.326	0.494 / 0.210	15.0 / 12.1	6.0 / 4.8	1.87 / 2.05	1.7 / 1.2	1.0 / 0.82	0.63 / 0.65
4 × 2⅝	9.5 / 7.7	2.76 / 2.21	4.00 / 4.00	2.796 / 2.660	0.293 / 0.293	0.326 / 0.190	6.7 / 6.0	3.3 / 3.0	1.56 / 1.64	0.91 / 0.77	0.65 / 0.58	0.58 / 0.59
3 × 2⅝	7.5 / 5.7	2.17 / 1.64	3.00 / 3.00	2.509 / 2.330	0.260 / 0.260	0.349 / 0.170	2.9 / 2.5	1.9 / 1.7	1.15 / 1.23	0.59 / 0.46	0.47 / 0.40	0.52 / 0.53

TABLE A-14
Properties of rolled-steel wide-flange shapes.[1]

nominal size in	weight lbf/ft	area in²	depth in	flange width in	t_f in	web t_w in	axis X-X I in⁴	I/c in³	k in	axis Y-Y I in⁴	I/c in³	k in
36 × 16½	300	88.17	36.72	16.655	1.680	0.945	20,290.2	1105.1	15.17	1225.2	147.1	3.73
	260	76.56	36.24	16.555	1.440	0.845	17,233.8	951.1	15.00	1020.6	123.3	3.65
	230	67.73	35.88	16.475	1.260	0.765	14,988.4	835.5	14.88	870.9	105.7	3.59
36 × 12	194	57.11	36.48	12.117	1.260	0.770	12,103.4	663.6	14.56	355.4	58.7	2.49
	170	49.98	36.16	12.027	1.100	0.680	10,470.0	579.1	14.47	300.6	50.0	2.45
	150	44.16	35.84	11.972	0.940	0.625	9,012.1	502.9	14.29	250.4	41.8	2.38
30 × 15	210	61.78	30.38	15.105	1.315	0.775	9,872.4	649.9	12.64	707.9	93.7	3.38
	190	55.90	30.12	15.040	1.185	0.710	8,825.9	586.1	12.57	624.6	83.1	3.34
	172	50.65	29.88	14.985	1.065	0.655	7,891.5	528.2	12.48	550.1	73.4	3.30
27 × 10	114	33.53	27.28	10.070	0.932	0.570	4,080.5	299.2	11.03	149.6	29.7	2.11
	102	30.01	27.07	10.018	0.827	0.518	3,604.1	266.3	10.96	129.5	25.9	2.08
24 × 14	160	47.04	24.72	14.091	1.135	0.656	5,110.3	413.5	10.42	492.6	69.9	3.23
	145	42.62	24.49	14.043	1.020	0.608	4,561.0	372.5	10.34	434.3	61.8	3.19
	130	38.21	24.25	14.000	0.900	0.565	4,009.5	330.7	10.24	375.2	53.6	3.13

24 × 9	94	27.63	24.29	9.061	0.872	0.516	2,683.0	220.9	9.85	102.2	22.6	1.92
	84	24.71	24.09	9.015	0.772	0.470	2,364.3	196.3	9.78	88.3	19.6	1.89
	76	22.37	23.91	8.985	0.682	0.440	2,096.4	175.4	9.68	76.5	17.0	1.85
18 × 11¾	114	33.51	18.48	11.833	0.991	0.595	2,033.8	220.1	7.79	255.6	43.2	2.76
	105	30.86	18.32	11.792	0.911	0.554	1,852.5	202.2	7.75	231.0	39.2	2.73
	96	28.22	18.16	11.750	0.831	0.512	1,674.7	184.4	7.70	206.8	35.2	2.71
18 × 8¾	85	24.97	18.32	8.838	0.911	0.526	1,429.9	156.1	7.57	99.4	22.5	2.00
	70	20.56	18.00	8.750	0.751	0.438	1,153.9	128.2	7.49	78.5	17.9	1.95
18 × 7½	60	17.64	18.25	7.558	0.695	0.416	984.0	107.8	7.47	47.1	12.5	1.63
	55	16.19	18.12	7.532	0.630	0.390	889.9	98.2	7.41	42.0	11.1	1.61
	50	14.71	18.00	7.500	0.570	0.358	800.6	89.0	7.38	37.2	9.9	1.59
16 × 8½	78	22.92	16.32	8.586	0.875	0.529	1,042.6	127.8	6.74	87.5	20.4	1.95
	64	18.80	16.00	8.500	0.715	0.443	833.8	104.2	6.66	68.4	16.1	1.91
16 × 7	50	14.70	16.25	7.073	0.628	0.380	655.4	80.7	6.68	34.8	9.8	1.54
	40	11.77	16.00	7.000	0.503	0.307	515.5	64.4	6.62	26.5	7.6	1.50
14 × 16	426	125.25	18.69	16.695	3.033	1.875	6,610.3	707.4	7.26	2,359.5	282.7	4.34
	370	108.78	17.94	16.475	2.658	1.655	5,454.2	608.1	7.08	1,986.0	241.1	4.27
	314	92.30	17.19	16.235	2.283	1.415	4,399.4	511.9	6.90	1,631.4	201.0	4.20
	264	77.63	16.50	16.025	1.938	1.205	3,526.0	427.4	6.74	1,331.9	166.1	4.14
	228	67.06	16.00	15.865	1.688	1.045	2,942.4	367.8	6.62	1,124.8	141.8	4.10
	202	59.39	15.63	15.750	1.503	0.930	2,538.8	324.9	6.54	979.7	124.4	4.06
	176	51.73	15.25	15.640	1.313	0.820	2,149.6	281.9	6.45	837.9	107.1	4.02
	150	44.08	14.88	15.515	1.128	0.695	1,786.9	240.2	6.37	702.5	90.6	3.99

[1] For additional sizes, see the current volume of "Steel Construction," American Institute of Steel Construction (AISC), New York.

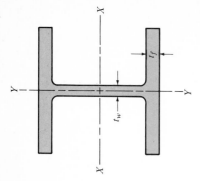

TABLE A-14
Properties of rolled-steel wide-flange shapes (continued). [1]

nominal size in	weight lbf/ft	area in²	depth in	flange width in	t_f in	web t_w in	axis X-X I in⁴	I/c in³	k in	axis Y-Y I in⁴	I/c in³	k in
14 × 14½	136	39.98	14.75	14.740	1.063	0.660	1593.0	216.0	6.31	567.7	77.0	3.77
	111	32.65	14.37	14.620	0.873	0.540	1266.5	176.3	6.23	454.9	62.2	3.73
	87	25.56	14.00	14.500	0.688	0.420	966.9	138.1	6.15	349.7	48.2	3.70
14 × 12	84	24.71	14.18	12.023	0.778	0.451	928.4	130.9	6.13	225.5	37.5	3.02
	78	22.94	14.06	12.000	0.718	0.428	851.2	121.1	6.09	206.9	34.5	3.00
14 × 10	74	21.76	14.19	10.072	0.783	0.450	796.8	112.3	6.05	133.5	26.5	2.48
	61	17.94	13.91	10.000	0.643	0.378	641.5	92.2	5.98	107.3	21.5	2.45
14 × 8	53	15.59	13.94	8.062	0.658	0.370	542.1	77.8	5.90	57.5	14.3	1.92
	48	14.11	13.81	8.031	0.593	0.339	484.9	70.2	5.86	51.3	12.8	1.91
	43	12.65	13.68	8.000	0.528	0.308	429.0	62.7	5.82	45.1	11.3	1.89
14 × 6¾	38	11.17	14.12	6.776	0.513	0.313	385.3	54.6	5.87	24.6	7.3	1.49
	30	8.81	13.86	6.733	0.383	0.270	289.6	41.8	5.73	17.5	5.2	1.41

APPENDICES *353*

Size	Wt											
12 × 12	190	55.86	14.38	12.670	1.736	1.060	1892.5	263.2	5.82	589.7	93.1	3.25
	120	35.31	13.12	12.320	1.106	0.710	1071.7	163.4	5.51	345.1	56.0	3.13
	92	27.06	12.62	12.155	0.856	0.545	788.9	125.0	5.40	256.4	42.2	3.08
	72	21.16	12.25	12.040	0.671	0.430	597.4	97.5	5.31	195.3	32.4	3.04
12 × 10	58	17.06	12.19	10.014	0.641	0.359	476.1	78.1	5.28	107.4	21.4	2.51
	53	15.59	12.06	10.000	0.576	0.345	426.2	70.7	5.23	96.1	19.2	2.48
12 × 8	50	14.71	12.19	8.077	0.641	0.371	394.5	64.7	5.18	56.4	14.0	1.96
	45	13.24	12.06	8.042	0.576	0.336	350.8	58.2	5.15	50.0	12.4	1.94
	40	11.77	11.94	8.000	0.516	0.294	310.1	51.9	5.13	44.1	11.0	1.94
10 × 10	112	32.92	11.38	10.415	1.248	0.755	718.7	126.3	4.67	235.4	45.2	2.67
	89	26.19	10.88	10.275	0.998	0.615	542.4	99.7	4.55	180.6	35.2	2.63
	72	21.18	10.50	10.170	0.808	0.510	420.7	80.1	4.46	141.8	27.9	2.59
	60	17.66	10.25	10.075	0.683	0.415	343.7	67.1	4.41	116.5	23.1	2.57
	49	14.40	10.00	10.000	0.558	0.340	272.9	54.6	4.35	93.0	18.6	2.54
10 × 8	45	13.24	10.12	8.022	0.618	0.350	248.6	49.1	4.33	53.2	13.3	2.00
	39	11.48	9.94	7.990	0.528	0.318	209.7	42.2	4.27	44.9	11.2	1.98
	33	9.71	9.75	7.964	0.433	0.292	170.9	35.0	4.20	36.5	9.2	1.94
10 × 5¾	29	8.53	10.22	5.799	0.500	0.289	157.3	30.8	4.29	15.2	5.2	1.34
	21	6.19	9.90	5.750	0.340	0.240	106.3	21.5	4.14	9.7	3.4	1.25
8 × 8	67	19.70	9.00	8.287	0.933	0.575	271.8	60.4	3.71	88.6	21.4	2.12
	48	14.11	8.50	8.117	0.683	0.405	183.7	43.2	3.61	60.9	15.0	2.08
	35	10.30	8.12	8.027	0.493	0.315	126.5	31.1	3.50	42.5	10.6	2.03
8 × 5¼	20	5.88	8.14	5.268	0.378	0.248	69.2	17.0	3.43	8.5	3.2	1.20
	17	5.00	8.00	5.250	0.308	0.230	56.4	14.1	3.36	6.7	2.6	1.16

¹ For additional sizes, see the current volume of "Steel Construction," American Institute of Steel Construction (AISC), New York.

TABLE A-15
Properties of sections.

A = area, in^2
I = rectangular area moment of inertia, in^4
J = polar area moment of inertia, in^4

$Z = \dfrac{I}{c}$ = section modulus, in^3

k = radius of gyration, in
$\bar{y}$ = centroidal distance, in

Rectangle

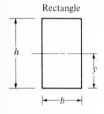

$$A = bh \qquad k = 0.289h$$

$$I = \frac{bh^3}{12} \qquad y = \frac{h}{2}$$

$$Z = \frac{bh^2}{6}$$

Triangle

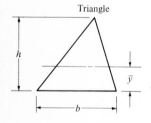

$$A = \frac{bh}{2} \qquad k = 0.236h$$

$$I = \frac{bh^3}{36} \qquad \bar{y} = \frac{h}{3}$$

$$Z = \frac{bh^2}{24}$$

Circle

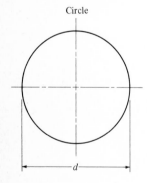

$$A = \frac{\pi d^2}{4} \qquad J = \frac{\pi d^4}{32}$$

$$I = \frac{\pi d^4}{64} \qquad k = \frac{d}{4}$$

$$Z = \frac{\pi d^3}{32} \qquad \bar{y} = \frac{d}{2}$$

Hollow circle

$$A = \frac{\pi}{4}\,(D^2 - d^2) \qquad J = \frac{\pi}{32}\,(D^4 - d^4)$$

$$I = \frac{\pi}{64}\,(D^4 - d^4) \qquad k = \frac{1}{4}\,\sqrt{D^2 + d^2}$$

$$Z = \frac{\pi}{32d}\,(D^4 - d^4) \qquad \bar{y} = \frac{D}{2}$$

TABLE A-15
Properties of sections (continued).

Ellipse

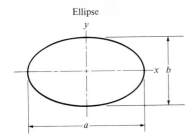

$$A = \frac{\pi ab}{4}$$

$$I_x = \frac{\pi ab^3}{64} \qquad I_y = \frac{\pi a^3 b}{64}$$

$$Z_x = \frac{\pi ab^2}{32} \qquad Z_y = \frac{\pi a^2 b}{32}$$

$$k_x = \frac{b}{4} \qquad k_y = \frac{a}{4}$$

Semicircle

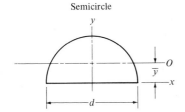

$$A = \frac{\pi d^2}{8} \qquad \bar{y} = \frac{2d}{3\pi}$$

$$I_x = \frac{\pi d^4}{128} \qquad I_o = 0.0069 d^4$$

$$k_x = \frac{d}{4} \qquad k_o = 0.132 d$$

TABLE A-16
Mechanical properties of some aluminum alloys.

These are *typical* properties for sizes of about $\frac{1}{2}$ in. A typical property may be neither a mean, a maximum, nor a minimum, but rather one that can be obtained provided reasonable care is taken in processing and in inspection. The values given for fatigue strength correspond to $50\,E+07$ cycles of completely reversed stress. The yield strengths are the 0.2 percent offset values. The shear modulus of rupture is tabulated as the shear ultimate strength.

material alloy number	temper	process-ing	strength tensile ultimate, kpsi	tensile yield, kpsi	shear ultimate, kpsi	compressive yield, kpsi	fatigue, kpsi	elongation in 2 in, percent	Brinell hardness number
1100	−O	wrought	13	5	9.5		5	45	23
	−H12	wrought	15.5	14	10		6	25	28
	−H14	wrought	22	20	14		9	16	40
	−H16	wrought	26	24	15		9.5	14	47
	−H18	wrought	29	27	16		10	10	55
3003	−O	wrought	16	6	11		7	40	28
	−H12	wrought	19	17	12		8	20	35
	−H14	wrought	22	20	14		9	16	40
	−H16	wrought	26	24	15		9.5	14	47
	−H18	wrought	29	27	16		10	10	55
3004	−O	wrought	26	10	16		14	25	45
	−H32	wrought	31	22	17		14.5	17	52
	−H34	wrought	34	27	18		15	12	63
	−H36	wrought	37	31	20		15.5	9	70
	−H38	wrought	40	34	21		16	6	77
2024	−O	wrought	27	11	18		13	22	47
	−T3	wrought	70	50	41		20	16	120
	−T4	wrought	68	48	41		20	19	120
	−T36	wrought	73	57	42		18	13	130
5052	−O	wrought	28	13	18		17	30	45
	−H32	wrought	34	27	20		17.5	18	62
	−H34	wrought	37	31	21		18	14	67
	−H36	wrought	39	34	23		18.5	10	74
	−H38	wrought	41	36	24		19	8	85
5056	−O	wrought	42	22	26		20	35	
	−H18	wrought	63	59	34		22	10	
	−H38	wrought	60	50	32		22	15	
43		sand cast	19	8	14	9	8	8	40
122	−T2	sand cast	27	20	21	20	9.5	1.0	80
142	−T21	sand cast	27	18	21	18	6.5	1.0	70
142	−T571	sand cast	32	30	26	34	8	0.5	85
212		sand cast	23	14	20	14	9	2.0	65
214		sand cast	25	12	20	12	7	9.0	50
220	−T4	sand cast	46	25	33	26	8	14	75

TABLE A-17
Mechanical properties of gray cast iron.

The American Society for Testing Materials (ASTM) numbering system for gray cast iron was devised so that the numbers correspond to the *minimum tensile strength* in kpsi. Thus, one can expect a tensile strength of at least 20,000 psi with an ASTM No. 20 cast iron; but note from the table that the typical value is 22,000 psi. The modulus of elasticity in both tension and torsion may vary as much as 10 percent in either direction from the tabulated values. Also, note that the modulus of rupture is called the *ultimate shear strength* in the table.

ASTM number	*strength*				*modulus of elasticity*		*Brinell hardness number*
	tensile, kpsi	*compressive, kpsi*	*shear ultimate, kpsi*	*endurance limit, kpsi*	*tension, Mpsi*	*torsion, Mpsi*	
20	22	83	26	10	12	4.8	156
25	26	97	32	11.5	13	5.3	174
30	31	109	40	14	15	5.9	201
35	36	124	48	16	16	6.4	212
40	42	140	57	18.5	18	7.1	235
50	52	164	73	21.5	21	7.6	262
60	62	187	88	24.5	22	8.2	302

TABLE A-18
Mechanical properties of some AISI steels.

The American Iron and Steel Institute (AISI) numbering system for steels is basically a four-number system in which the first two numbers refer to the alloy composition (aside from carbon) and the last two numbers give the approximate carbon content. The properties shown below have been obtained from a variety of sources and are believed to be representative. Under processing, *HR* means hot rolled, *CD* means cold drawn or cold finished, and the designation drawn 800°F, say, means the sample was heat treated and then tempered (drawn) to a temperature of 800°F. The strengths shown are for sizes in the $\frac{3}{4}$ to 1-in range.

AISI number	carbon content, percent	alloy composition	processing	tensile strength, kpsi	yield strength, kpsi	elongation in 2 in, percent	Brinell hardness number
1010	0.08–0.13		HR	47	26	28	95
			CD	53	44	20	105
1015	0.13–0.18		HR	50	27	28	101
			CD	56	47	18	111
1018	0.15–0.20		HR	58	32	25	116
			CD	64	54	15	126
1112	0.13 max		HR	56	33	25	121
			CD	78	60	10	167
1035	0.32–0.38		HR	72	39	18	143
			CD	80	67	12	163
			Drawn 800°F	110	81	18	220
			Drawn 1000°F	103	72	23	201
			Drawn 1200°F	91	62	27	180
1040	0.37–0.44		HR	76	42	18	149
			CD	85	71	12	170
			Drawn 1000°F	113	86	23	235
1045	0.43–0.50		HR	82	45	16	163
			CD	91	77	12	179
1050	0.48–0.55		HR	90	49	15	179
			CD	100	84	10	197
			Drawn 600°F	220	180	10	450
			Drawn 900°F	155	130	18	310
			Drawn 1200°F	105	80	28	210
2330	0.28–0.34	Ni	Drawn 400°F	221	195	11	425
			Drawn 600°F	196	171	14	382
			Drawn 800°F	160	131	18	327
			Drawn 1000°F	127	97	23	268
			Drawn 1200°F	108	70	27	222
3140	0.38–0.43	Ni-Cr	HR	96	64	26	197
			CD	104	91	17	212
			Drawn 800°F	188	157	15	376

TABLE A-18
Mechanical properties of some AISI steels (continued).

AISI number	carbon content, percent	alloy composition	processing	tensile strength, kpsi	yield strength, kpsi	elongation in 2 in, percent	Brinell hardness number
4130	0.28–0.43	Cr-Mo	HR	90	60	30	183
			CD	98	87	21	201
			Drawn 1000°F	146	133	17	293
4140	0.38–0.43	Cr-Mo	HR	90	63	27	187
			CD	102	90	18	223
			Drawn 1000°F	153	131	16	302
4340	0.38–0.43	Ni-Cr-Mo	HR	101	69	21	207
			CD	111	99	16	223
			Drawn 600°F	260	234	12	498
			Drawn 1000°F	182	162	15	363

TABLE A-19
ASTM mechanical property specifications for hot-rolled steel.[1]

ASTM No.	grade	subject	*range of tensile strength*		
			bars, kpsi	*plate,* kpsi	*shapes,* kpsi
A7		Bridges and buildings	60–72	60–75	60–72
A31	A	Boiler rivet steel	45–55		
	B		58–68		
A94		Structural silicon steel		80–95	80–95
A113	A	Locomotives and cars	60–72		60–72
	B		50–62		50–62
	C		48–58		48–58
A131		Ships			58–71
		Ships, rivet steel	55–65		
A141		Structural rivet steel	52–62		
A195		Rivet steel	68–82		
A201	A	Boilers, welded		55–65	
	B			60–72	
A212	A	Boilers		65–77	
	B			70–85	
A283	A	Structural quality		45–55	
	B			50–60	
	C			55–65	
	D			60–72	
A284	A	Machine parts		50 min	
	B			55 min	
	C			60 min	
A285	A	Flange and firebox		45–55	
	B			50–60	
	C			55–65	
A299		Boilers		75–90	
A306	45	Structural-general	45–55		
	50		50–60		
	55		55–65		
	60		60–72		
	65		65–77		
	70		70–85		
	75		75–90		
	80		80 min		
A373		Structural, welded			58–75

[1] See "Metals Handbook," 8th ed., American Society for Metals, Metals Park, O., 1961, p. 70.

TABLE A-20
Deflection formulas for beams.

1. Cantilever with end load.

$$y = \frac{Fx^2}{6EI}(x - 3l) \qquad y_{max} = -\frac{Fl^3}{3EI}$$

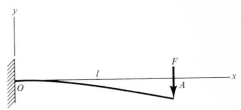

2. Cantilever with intermediate load.

$$y_{OA} = \frac{Fx^2}{6EI}(x - 3a) \qquad y_{AB} = \frac{Fa^2}{6EI}(a - 3x)$$

$$y_{max} = \frac{Fa^2}{6EI}(a - 3l)$$

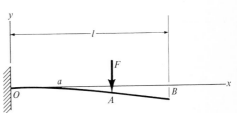

3. Cantilever with uniform load.

$$y = \frac{wx^2}{24EI}(4lx - x^2 - 6l^2) \qquad y_{max} = -\frac{wl^4}{8EI}$$

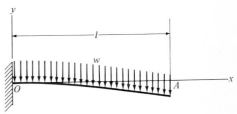

4. Cantilever with moment load.

$$y = \frac{M_A x^2}{2EI} \qquad y_{max} = \frac{M_A l^2}{2EI}$$

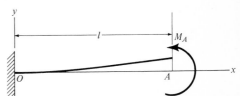

5. Simply supported with center load.

$$y_{OA} = \frac{Fx}{48EI}(4x^2 - 3l^2) \qquad y_{max} = -\frac{Fl^3}{48EI}$$

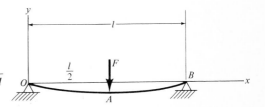

6. Simply supported with intermediate load.

$$y_{OA} = \frac{Fbx}{6EIl}(x^2 + b^2 - l^2)$$

$$y_{AB} = \frac{Fa(l - x)}{6EIl}(x^2 + a^2 - 2lx)$$

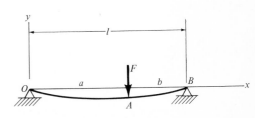

TABLE A-20
Deflection formulas for beams (continued).

7. Simply supported with uniform load.

$$y = \frac{wx}{24EI}\,(2lx^2 - x^3 - l^3)$$

$$y_{max} = -\frac{5wl^4}{384EI}$$

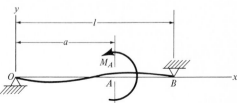

8. Simply supported with moment load.

$$y_{OA} = \frac{M_A x}{6EIl}\,(x^2 + 3a^2 - 6al + 2l^2)$$

$$y_{AB} = \frac{M_A}{6EIl}\,[x^3 - 3lx^2 + x(2l^2 + 3a^2) - 3a^2 l]$$

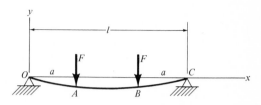

9. Simply supported with twin loads.

$$y_{OA} = \frac{Fx}{6EI}\,(x^2 + 3a^2 - 3la)$$

$$y_{AB} = \frac{Fa}{6EI}\,(3x^2 + a^2 - 3lx)$$

$$y_{max} = \frac{Fa}{24EI}\,(4a^2 - 3l^2)$$

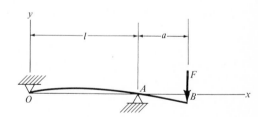

10. Simply supported with overhanging load.

$$y_{OA} = \frac{Fax}{6EIl}\,(l^2 - x^2)$$

$$y_{AB} = \frac{F(x-l)}{6EI}\,[(x-l)^2 - a(3x-l)]$$

$$y_B = -\frac{Fa^2}{3EI}\,(l+a)$$

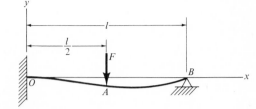

11. One fixed and one simple support with center load.

$$y_{OA} = \frac{Fx^2}{96EI}\,(11x - 9l)$$

$$y_{AB} = \frac{F(l-x)}{96EI}\,(5x^2 + 2l^2 - 10lx)$$

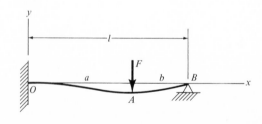

12. One fixed and one simple support with intermediate load.

$$y_{OA} = \frac{Fbx^2}{12EIl^3}\,[3l(b^2 - l^2) + x(3l^2 - b^2)]$$

$$y_{AB} = y_{OA} - \frac{F(x-a)^3}{6EI}$$

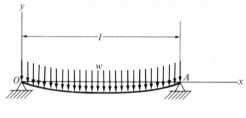

13. One fixed and one simple support with uniform load.

$$y = \frac{wx^2}{48EI} (l - x)(2x - 3l)$$

$$y_{max} = -\frac{wl^4}{185EI}$$

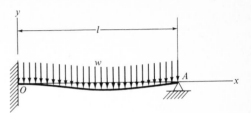

14. Both ends fixed with center load.

$$y_{OA} = \frac{Fx^2}{48EI} (4x - 3l) \qquad y_{max} = -\frac{Fl^3}{192EI}$$

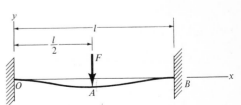

15. Both ends fixed with intermediate load.

$$y_{OA} = \frac{Fb^2x^2}{6EIl^3} [x(3a + b) - 3al]$$

$$y_{AB} = \frac{Fa^2(l - x)^2}{6EIl^3} [(l - x)(3b + a) - 3bl]$$

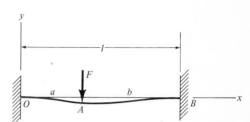

16. Both ends fixed with uniform load.

$$y = -\frac{wx^2}{24EI} (l - x)^2 \qquad y_{max} = -\frac{wl^4}{384EI}$$

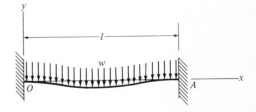

index